D1267015

TESLA

THE MAN, THE INVENTOR AND THE AGE OF ELECTRICITY

DR RICHARD
GUNDERMAN

ANDRE
DEUTSCH

Published in 2019 by André Deutsch
An imprint of the Carlton Publishing Group
20 Mortimer Street
London W1T 3JW

10 9 8 7 6 5 4 3 2

A CIP catalogue record for this book is available from the British Library.

ISBN 978 0 233 00576 8

Printed in Dubai

CONTENTS

Preface and Acknowledgements

Tesla's is a truly Promethean tale. Prometheus, whose name means forethought, was both a Titan of Greek mythology, and a trickster who is credited with creating mankind from clay and stealing fire from the gods and giving it to humanity. In punishment, Zeus chained Prometheus to a rock, where an eagle would come and eat his liver (the organ of life), which would then grow back overnight, the cycle being repeated each day.

Like Prometheus, Nikola Tesla was always looking to the future. He sought to wrench lightning from the heavens and harness it to man's will, offering what he regarded as some of the greatest benefactions in the history of humankind. He, too, relished the role of wonder worker, amazing his audiences with seemingly supernatural demonstrations that often eclipsed his scientific and technical work. And in true Promethean fashion, he spent the last decades of his life alone and tormented by loss.

As a practising physician, I first encountered Tesla as the unit measure of magnetic flux density of MRI scanners. One day, curious to know more about the man, I did some digging. To my delight, his life story turned out to be more remarkable than I imagined. Yet when I approached my colleagues, few were familiar with it. How could such an incredible tale go untold? So, some years ago, a medical student, Aleks Alavanja, and I drafted a manuscript on Tesla for a radiology journal.

As I dug more deeply, however, it became apparent that Tesla's story would hold broad appeal for a general audience. Weaving together where he came from, what he set out to do, what he actually managed to accomplish, the time in which he lived, the people with whom he worked, the great fame he achieved, and the many setbacks and sufferings he endured, would yield a tapestry that no curious person could fail to find fascinating. But how to tell the story?

The hand of fate seemed to be at work when I received a message from Isabel Wilkinson at André Deutsch. She had seen a piece of mine in *The Conversation* celebrating Tesla's life on the seventy-fifth anniversary of his death, and she inquired whether I might have interest in a richer telling of the story in book form. Mindful of my colleagues and friends who would both relish and find insight in Tesla's tale, I leapt at the opportunity, distilling the fruits of years of sleuthing into 60 days of writing.

For their patience with my immersion in this project, I would like to thank my children Rebecca, Peter, and David, and especially my son John and wife Laura, who bore the brunt of the manuscript's birth; my many colleagues at Indiana University in the schools of medicine, liberal arts, and philanthropy, who provided opportunities to pursue this passion; and my teachers at Wabash College and the University of Chicago, who planted a number of seeds that germinate here.

Dr Richard Gunderman

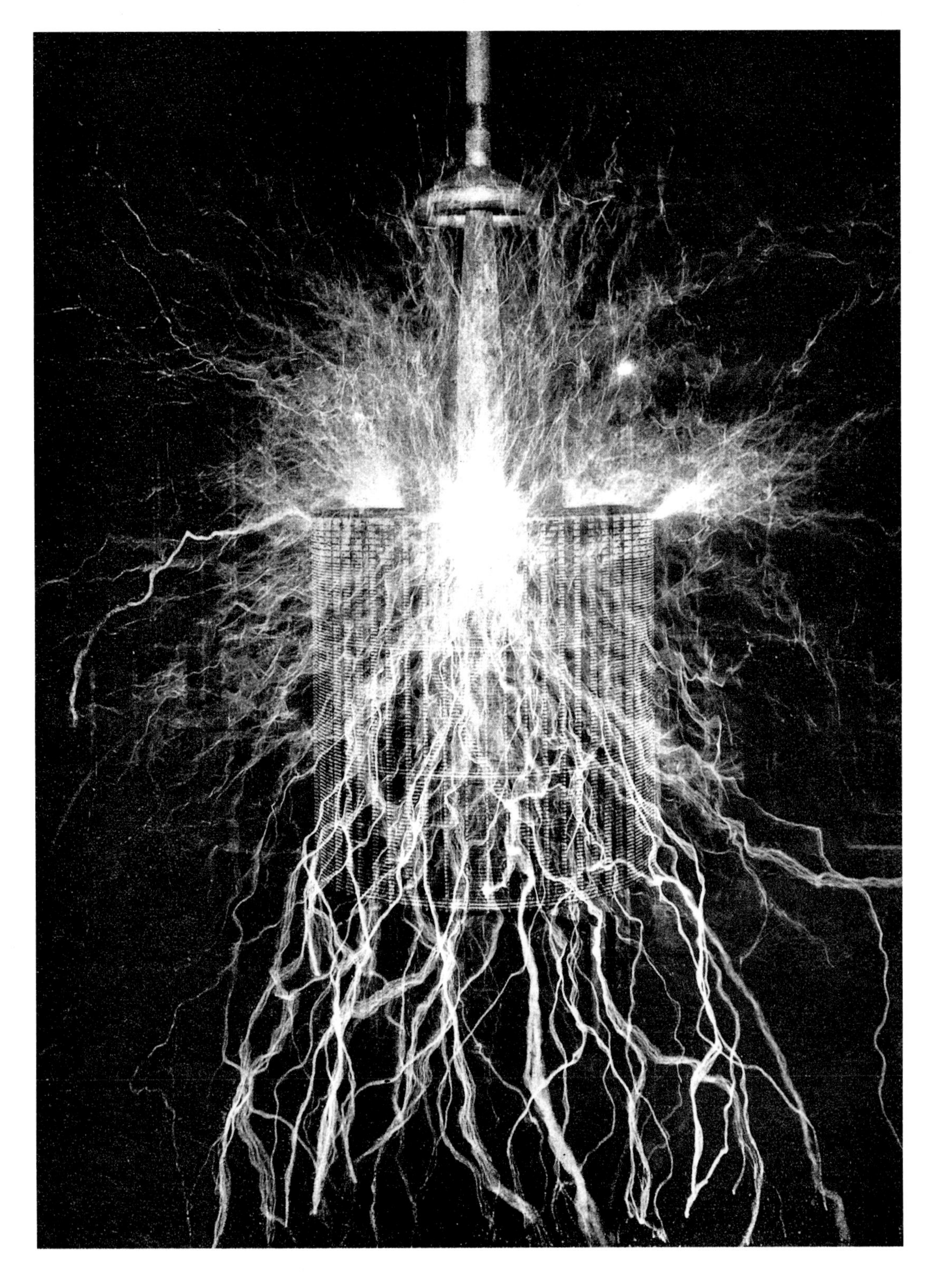

TIMELINE

1856

Tesla born to a Serbian family in what is now Croatia

1863

Tesla's older brother Dane killed in horse riding accident

1870

Tesla's studies awaken a life-long fascination with physics

1873

Tesla contracts cholera and extracts from his father a pledge to allow him to attend engineering school

1874

After his recovery, Tesla roams the mountains, perhaps in part to avoid military service

1875

Tesla excels as an engineering student in Graz, though within two years he would drop out

1879

Tesla's father dies

1881

Tesla goes to work for the telephone company in Budapest, where he becomes chief electrician

1882

After a vision of the AC motor comes to Tesla in a flash, he moves to Paris to work for Continental Edison

1884

With funding from his uncles, Tesla sails to New York, where he goes to work for Edison

1885

The Tesla Electric Light Company is formed in New Jersey, though Tesla is soon forced out

1887

After working as a ditch digger, Tesla and investors form the Tesla Electric Company

1888

Tesla sells his AC Polyphase system to Westinghouse

1890

Tesla agrees to forfeit AC royalties from Westinghouse

1891

Tesla becomes naturalized citizen of the US and patents the Tesla coil

1893

Tesla's Polyphase AC system lights the World's Columbian Exposition, where Tesla exhibits many electrical effects

1895

Tesla and investors form the Nikola Tesla Company, which would handle his patents for decades

1896

Tesla-inspired generators begin sending Power from Niagara Falls to Buffalo, NY

1898

Tesla demonstrates his radio-controlled boat at Madison Square Garden

1899

Tesla sets up his experimental station in Colorado Springs

1900

Tesla announces that he has received signals that may be emanating from another planet

1901

With a $150,000 investment from JP Morgan, Tesla begins constructing his wireless transmission station at Wardenclyffe

1905

Lacking funds, Tesla ceases work at Wardenclyffe

1906

Tesla demonstrates his bladeless turbine

1909

Marconi wins the Nobel Prize for radio

1914

George Westinghouse dies

1915

Newspapers erroneously report that Tesla and Edison will share the Nobel Prize

1916

Tesla's desperate finances become public

1917

Tesla receives the Edison Medal

1919

Tesla begins publishing a series of autobiographical articles, "My Inventions"

1922

Tesla's beloved pigeon dies

1931

Tesla turns 75, receiving letters from many prominent scientists and engineers

1943

Tesla dies in New York at age 86

THE WORLD BEFORE ELECTRIFICATION

When the US National Academy of Engineering ranked the 20 greatest engineering achievements of the twentieth century, the list was topped by electrification. Not only did electrification crown the list, but many of the other top achievements – electronics, radio and television, computers, telephones, air conditioning and refrigeration, the internet and household appliances – were made possible by widely available electrical power. Since most contemporary readers grew up in an electrified world, it can be difficult to imagine what life was like before flipping a switch provided electrical power.

NAE GREATEST ENGINEERING ACHIEVEMENTS OF THE 20TH CENTURY

1. Electricification
2. Automobile
3. Airplane
4. Water Supply & Distribution
5. Electronics
6. Radio & Television
7. Agricultural Mechanization
8. Computers
9. Telephone
10. Air Conditioning & Refrigeration
11. Highways
12. Spacecraft
13. Internet
14. Imaging
15. Household Appliances
16. Health Technologies
17. Petroleum & Petrochemical Technologies
18. Laser & Fibre Optics
19. Nuclear Technologies
20. High-performance Materials

Consider the following passage from Daniel Defoe's 1719 novel *Robinson Crusoe*.

I was at a great loss for candles; so that as soon as ever it was dark, which was generally by seven o'clock, I was obliged to go to bed. I remembered the lump of beeswax with which I made candles in my African adventure; but I had none of that now; the only remedy I had was, that when I had killed a goat I saved the tallow, and with a little dish made of clay, which I baked in the sun, to which I added a wick of some oakum, I made me a lamp; and this gave me light.

As Crusoe's story indicates, throughout most of human history, the setting of the sun brought about the cessation of light-requiring activities. Reading, writing, and working on crafts were all but impossible, and even moving about safely in the dark could be quite difficult, particularly on cloudy or moonless nights. Fire provided some means of illumination, but as anyone who has attempted to read by firelight knows, the light is generally quite dim. As a result, farmers and other labourers often arose before daybreak and laboured till sunset, to ensure that no sunlight went to waste.

Over time, human beings developed means of lighting the night sky. The tallow that Crusoe refers to was made by boiling animal fat, rendering tallow candle production a smelly and messy business, requiring the handling of animal carcasses. When burnt, such candles tended to produce a good deal of smoke. Beeswax burned more cleanly but was too expensive for ordinary people. The Great American polymath Benjamin Franklin (1706–90) was the son of a tallow chandler whose modest means prevented him from providing more than two years of formal schooling for his son. Over time, tallow candles were replaced by oil from whale blubber and later kerosene.

Of course, light was only one aspect of life revolutionized by electrification. In the pre-electric world, many foods spoiled quickly, and meats had to be smoked or salted. Cooking generally required fire, powered by wood, charcoal or gas. Heating a home also tended to involve fire, and fireplaces and stoves were the norm in colder climates. Naturally,

ABOVE. The 1719 (first) edition of Daniel Defoe's *Robinson Crusoe*, whose protagonist had to recreate for himself many of the conveniences of civilized life.

RIGHT. Pastel study of Benjamin Franklin (Joseph Duplessis, 1770s). Printer, entrepreneur, inventor, diplomat and statesman, Franklin's early investigations of electricity made him one of the most famous scientists of his day.

OPPOSITE. The National Academy of Engineering's 20 greatest engineering achievements of the twentieth century, which places electrification at number one. Note that many other items on the list presuppose the wide availability of electrical power.

the near-constant presence of fire in dwellings carried a dramatically increased risk of house fires. Keeping clothing clean generally involved a good deal of manual labour. Water for drinking, cooking and washing had to be hauled from a stream or pulled up from a well. Today electric appliances often do much of this work.

Although we speak of electrification as a twentieth century achievement, it is important to remember than over one billion people today lack electrical power in their homes. Such individuals are most highly concentrated in Africa and the Indian subcontinent. Not surprisingly, electrification and technological progress are closely connected. While most electrical power around the world is produced by steam turbines and delivered through large power grids, it is also possible to supply electricity to populations that lack coal, natural gas, or nuclear power plants through such means as hydroelectricity, wind turbines and solar energy, which can help make electricity available in more remote areas.

ACCESS TO ELECTRICITY (% POPULATION)

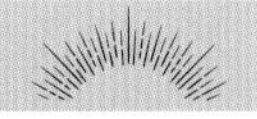

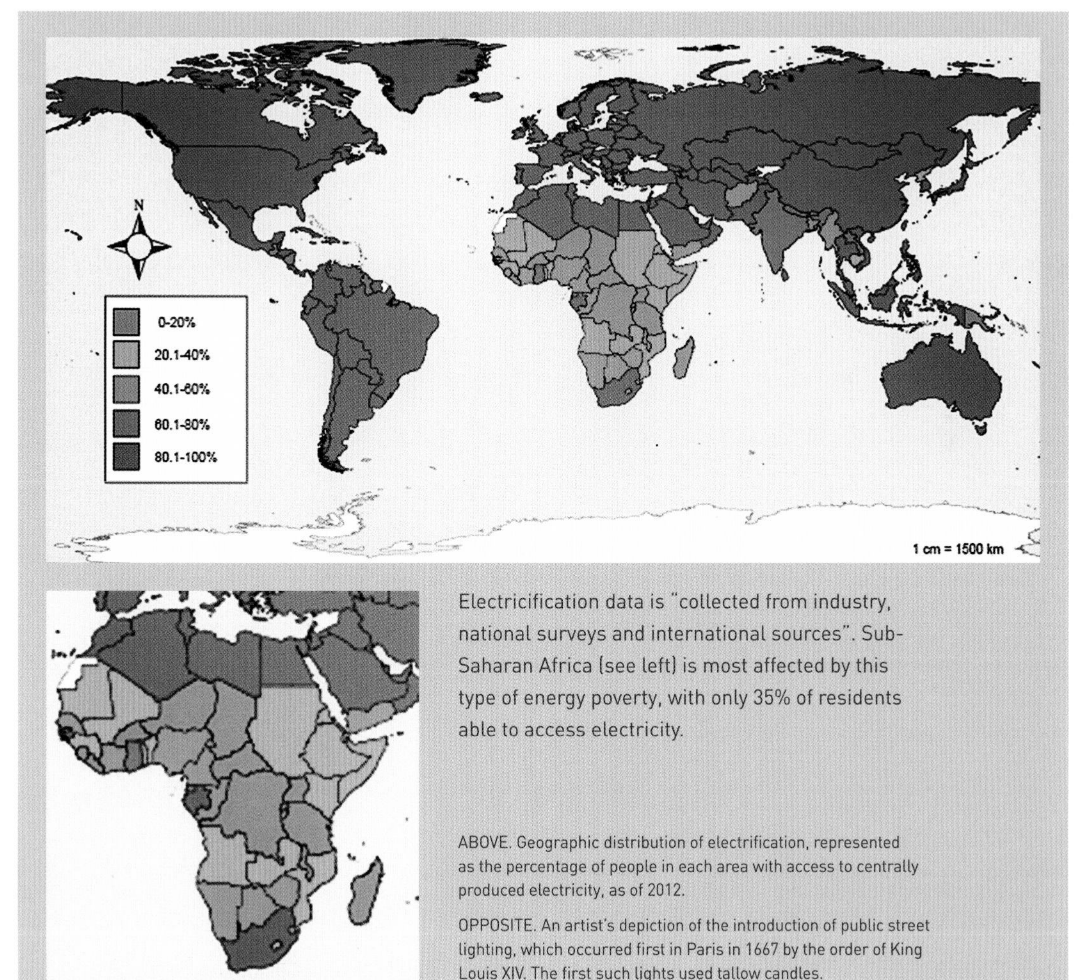

Electricification data is "collected from industry, national surveys and international sources". Sub-Saharan Africa (see left) is most affected by this type of energy poverty, with only 35% of residents able to access electricity.

ABOVE. Geographic distribution of electrification, represented as the percentage of people in each area with access to centrally produced electricity, as of 2012.

OPPOSITE. An artist's depiction of the introduction of public street lighting, which occurred first in Paris in 1667 by the order of King Louis XIV. The first such lights used tallow candles.

THE ELECTRICAL PIONEERS

Electricity has been around a lot longer than human beings, but human beings have been writing for thousands of years about electrical phenomena such as lightning, electric fish and static electricity. However, the unified concept of electricity only began to emerge around the year 1600. The word "electric" means "from amber", because it was known that rods of this substance rubbed with the fur of an animal would attract objects like feathers. Still more time would elapse before Italian scientist Luigi Galvani discovered the role of electricity in living organisms, inspiring Mary Shelley's novel *Frankenstein*, in which electricity makes possible the reanimation of dead body parts.

One of the first scientists to study electricity was Benjamin Franklin, the great statesman who was also one of the most famous scientists of his day, thanks to his discoveries concerning electricity. Franklin likened electricity to a fluid, because it could pass from one body to another. He also introduced the notion of positive and negative charges. Most famously, Franklin may have flown a kite in a thunderstorm, attaching it to a hemp string, which was in turn attached to a key. As the story goes, when Franklin saw strands of the hemp string stand erect, he put his finger near the key and felt a spark, proving that electricity was at work.

LEFT. The oldest extant illustration of Mary Shelley's *Frankenstein*, by Theodor von Holst (from the 1831 edition), depicting Victor Frankenstein fleeing in horror just after he reanimated the parts of dead bodies he had assembled into an eight-foot-tall creature.

MICHAEL FARADAY

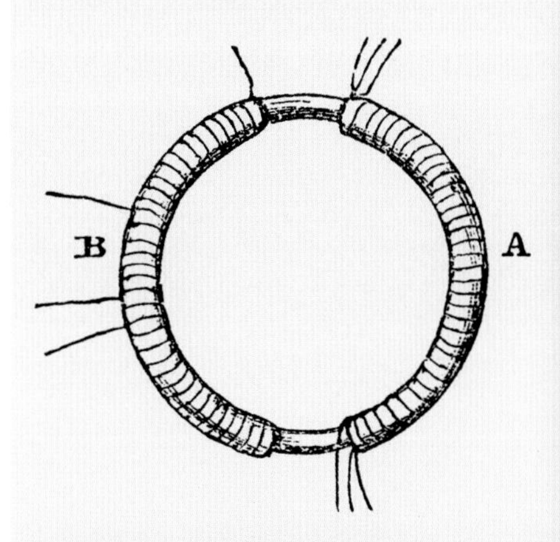

Perhaps the greatest of all scientific students of electricity was Michael Faraday (1791–1867), the great English experimentalist. Though lacking in formal education, Faraday made many seminal contributions in chemistry, such as the discovery of benzene. Yet it was his work on electromagnetism that sealed his reputation as one of the greatest scientists in history. He became the first person to produce an electric current from a magnetic field, invented the first electric motor, discovered that magnetic fields affect light, identified diamagnetism (the altered behaviour of some substances in strong magnetic fields), and showed that electricity is involved in chemical bonding.

ABOVE. An engraving based on a sketch in Faraday's notebook, showing an early electrical transformer. Current in one of the coil's wires induces current in the other.

ABOVE RIGHT. Michael Faraday, the greatest scientific investigator of electricity in human history, as portrayed in 1842 by Thomas Phillips.

RIGHT. Benjamin West's 1816 depiction of Franklin's famous kite experiment.

TESLA'S EARLY YEARS

Nikola Tesla was born an ethnic Serb in what is now Croatia, then part of the Austro-Hungarian empire. His family name, Tesla, refers to a type of axe. His father, Milutin, came from a military family but withdrew from the military academy and instead studied for the priesthood. At the time of Tesla's birth, his father was serving the parish of Smiljan, where he also found time to publish articles promoting such causes as education for all children and the equality of all peoples. His articles brought him to the attention of those beyond the village, including the prominent family of a priest with seven children, the eldest of whom was a daughter named Djuka, who became his wife.

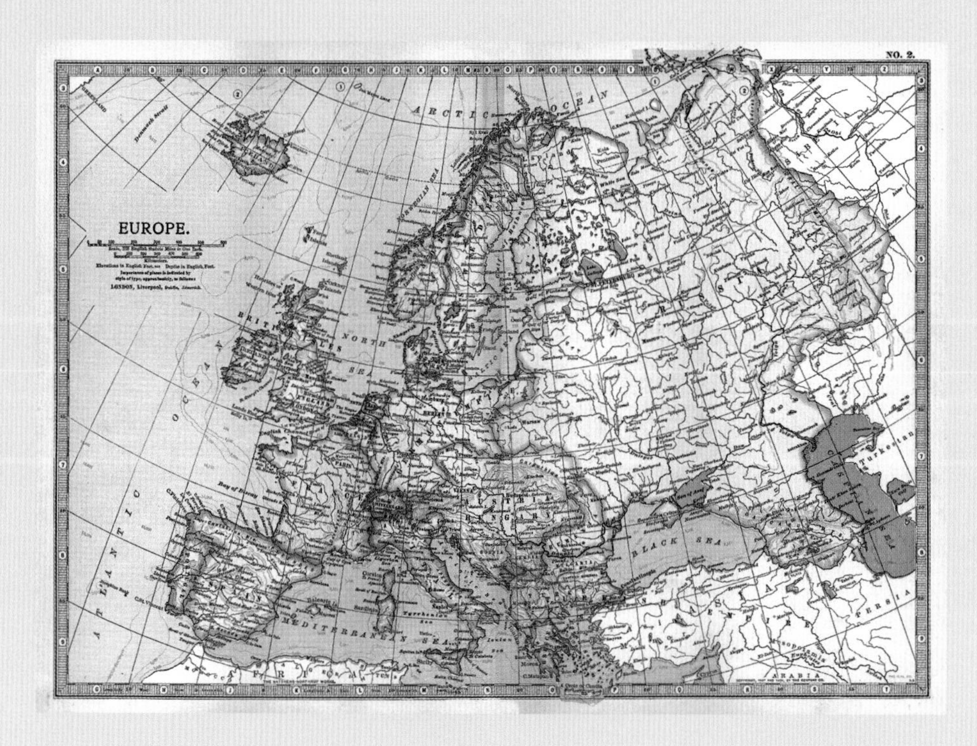

According to Tesla, the night of his birth was a stormy one. After a difficult labour, he was born at midnight between July 9 and 10, 1856, the fourth of his parents' five children. The midwife, exasperated by the difficult delivery, predicted gloomily that Tesla would be "a child of the dark". Just as he emerged, however, a flash of lightning tore across the sky, prompting his mother to counter, "No, of the light."

Tesla traced his extraordinary inventive abilities to his mother. When Djuka's mother became blind, she assumed responsibility for raising her younger siblings. Though illiterate, she could recite epic poems and long passages from the Bible. After marriage, she bore five children, of which Nikola was the fourth. She tended the farm, sewed her family's clothes, raised her children, and still found time to invent numerous household devices, including churns and looms. Tesla later wrote of her:

She would have achieved great things had she not been so remote from modern life and its multiple opportunities. She invented and constructed all kinds of tools and devices and wove the finest designs from thread which was spun by her. She even planted the seeds, raised the plants, and separated the fibres herself. She worked indefatigably, from break of day till late at night, and most of the wearing apparel and furnishings of the home were the product of her hands.

Tesla's parents regarded his elder brother Dane as the most gifted of their children. To make the most of his sons' intellects, their father, who spoke multiple

ABOVE. Photograph of Tesla's birthplace (left), next to an orthodox church (right), in the village of Smiljan.

OPPOSITE. Croatia's position on a modern map of Europe.

languages and could also recite long literary and Biblical passages from memory, challenged them with daily exercises. These included "guessing one another's thoughts, discovering the defects of some form of expression, repeating long sentences, or performing mental calculations." Nikola developed such a love of reading that his father would deprive him of candles so that he would not ruin his eyes by reading in dim light, but the boy soon began making his own so that he could read through the night.

Disaster struck the family when Dane was killed in an accident. His brother was riding the family's prize horse when he was thrown from his mount. Tesla reported that the death left his parents disconsolate:

> *The recollection of his attainments made every effort of mine seem dull in comparison. Anything that I did that was creditable merely caused my parents to feel their loss more keenly. So I grew up with little confidence in myself.*

Tesla's confidence was later revived during a demonstration of his town's new fire engine. When no water emerged from the hose, Tesla leapt into the river and unkinked the hose, restoring the flow of water and garnering the acclaim of those assembled.

From a young age, Tesla showed a penchant for invention. He built a waterwheel and a propeller powered by cockchafer beetles, and he even tried to fly from the roof of the family's barn, borne aloft (or so he thought) by an umbrella. The accident left him unconscious, but he recovered fully. He reported that as a child he first observed electrical phenomena while stroking the family cat. Seeing the animal's fur

ABOVE. Photographic portrait of Milutin, Tesla's father.

RIGHT. Matthew Brady's 1871 photographic portrait of Mark Twain.

OPPOSITE. Photograph of Tesla's boyhood home in the town of Gospic.

become a "sheet of light" and emit a shower of sparks, Tesla asked his father what it could be. "What you are seeing is nothing but electricity," his father replied. His mother, however, told him to stop, for fear that he might start a fire.

Tesla reported that as a child and adolescent, he developed multiple unusual phobias and fixations. He could not bear to touch another person's hair. He found the sight of pearls on a woman unbearable, and he later sent workers home for the day when he saw them so attired. He calculated in his head the amount of food on his plate and the volume of liquid in his cup, and he made a habit of counting the number of steps he had taken when walking. He also saw things so vividly in his imagination that he sometimes needed to ask his sisters whether what he was seeing was real or not. This faculty, which seemed a burden to him as a child, later played an important role in his career as an inventor.

Tesla was a sickly child, frequently bedridden by illness. A voracious reader, he was delighted to be given the job of cataloguing the books at the local library. While there, he claimed to have discovered the works of the great American writer Mark Twain, which Tesla credited with restoring his vigour. Twenty-five years later when he met Twain in person, Tesla shared with Twain the difference his writings had made in the inventor's life. According to Tesla, upon hearing this account, Twain's eyes filled with tears, he was so touched by the account. Twain would later describe Tesla's induction motor as "the most valuable patent since the telephone".

ABOVE LEFT. Photograph of Tesla's sister Marica.

ABOVE. Photograph of Tesla's sister Angelina.

LEFT. Photograph of Tesla's sister, Milka.

OPPOSITE. Nikola Tesla's birth certificate.

Изводъ

Протокола Крещаемыхъ, при восточной православной церкви, храма святыхъ Апостолъ, Петра и Павла, въ Смилянъ селѣ.

25 kr. 25 kr.

Ч. 21.

Родися младенецъ полъ мужескагω, мѣсяца іюніа, дне 28. лѣта 1856. /: пятьдесять и шестагω/ законнω. Отецъ младенца, Милутинъ Тесла, парохіи администраторъ, мати же Георгіна, жители Смилянскіи. Крестися и мѵромъ святымъ помазася чрезъ мене іерея Тому ω Окльотскія, парохіи Администратора церкве храма святагω великомученика и побѣдоносца Георгіа въ Госпићи селѣ, мѣсяца іюніа, дне 29. лѣта 1856. И дано бысть во святомъ крещеніи младенцу имя „Николай". Воспріемникъ бысть, Іωаннъ Дреновацъ, з.к. сотникъ, житель Госпићкій.

Да сей изводъ протокола крещаемыхъ, своему оригиналу со всѣмъ сходенъ есть, собственоручною подписію и приложеніемъ обычнаа печати свидѣтелствую.

Дано въ Госпићи /: Окружіе Лико-Оточкое:/ дне 19/31. октоврія 1883.

СРПСКЕ ...

Протоіерей, Петръ Мандић, парохъ Госпићкій. — М

Die Echtheit der Schrift u. Unterschrift des Herrn Pfarrers u. Erzpriesters Petar Mandić, dann des beigedrückten Kirchensiegels, wird hiemit bestätigt. Gospić, den 1/13. November 1883.

Der Gemeindevorstand,

OBĆINSKI URED U GOSPIĆU

Vinović

EDUCATION

Tesla's father expected his only surviving son to follow him into the priesthood, but Nikola's early studies convinced him that engineering was his passion. He had such an aptitude for it, as demonstrated by his ability to perform complex calculations in his head, that some of his teachers accused him of cheating. When Tesla contracted cholera at age 17, an illness that left him virtually bedridden – and, at points over a period of nine months, near death – he seized the occasion to extract from his father a promise that, if he recovered, Milutin would allow him to shift his course of study from the clergy to engineering. Tesla recovered soon thereafter, and his father fulfilled his promise.

Perhaps to avoid military service, Tesla then spent a year in the woods camping and hiking, ostensibly to regain his health. At the end of 1875, he enrolled at the engineering school at Graz, where he demonstrated extreme dedication to his studies, working from 3 am to 11 pm seven days per week. In his first year, he earned the highest possible grades in all his classes and completed twice the number of examinations expected of a first-year student. When he returned home for the summer, however, he was stunned when his father tried to talk him out of returning to school. Unbeknownst to Tesla, his teachers had written to Milutin, warning that the young man's study habits might kill him.

Yet Tesla did return, and in his second year one of his professors showed him a new direct-current dynamo that could be used as both a generator and a motor. Tesla reports that he saw almost immediately that the design rested on an inefficient foundation – namely, a component known as a commutator that regularly reversed the direction of the current between the rotor and the external circuit. Tesla envisioned that the commutator, which generated many sparks and needed to be replaced often, might be dispensed with. When he shared his intuition, however, his instructor ridiculed the suggestion, likening Tesla's vision to a perpetual motion machine. Tesla became obsessed with realizing his dream.

Tesla's academic fortunes began to wane. Changes in the budget of the military led to the loss of his fellowship, without which it would be nearly impossible for the son of a priest to continue his studies. Partly to support himself, Tesla turned to playing cards and billiards for money. Initially he did well, but eventually he fell into a losing streak that left him penniless. When he turned to his parents, Milutin was outraged that his son was gambling, and warned him to stop. Tesla refused. Later, his mother handed him a large wad of bills, urging him to go and gamble it away, saying, "The sooner you lose all we possess, the better it will be." Tesla abruptly abandoned his passion for gambling.

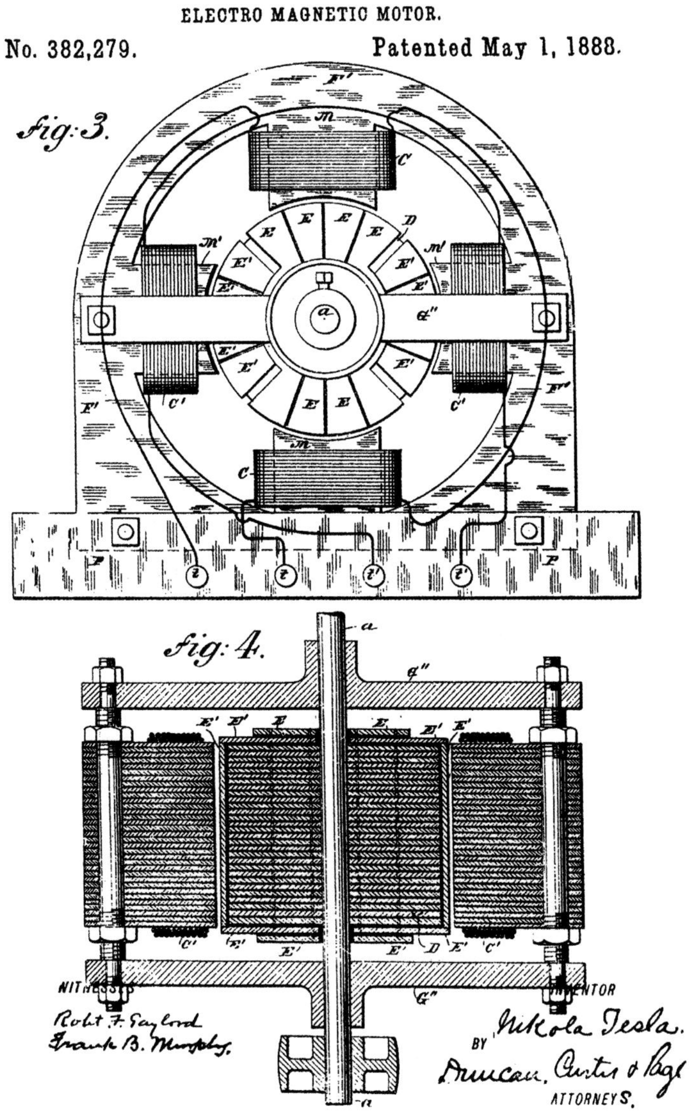

Tesla left the engineering school at Graz without a degree. He wandered from town to town, working at odd jobs before finally returning home to face his family. There Milutin soon passed away, leaving Tesla with a charge to continue his education. He went to study in Prague, where he encountered several remarkable faculty members who stimulated his mind in areas outside of engineering, including physics and philosophy. Although he once again left without a degree, Tesla later reported that it was during his months of study in Prague that he began to make real progress on the problem of generating electricity without the use of a commutator.

ELECTRO MAGNETIC MOTOR.

No. 382,279. Patented May 1, 1888.

Fig: 3.

Fig: 4.

WITNESSES
Robt. F. Gaylord
Frank B. Murphy.

INVENTOR
Nikola Tesla.
BY Duncan, Curtis & Page
ATTORNEYS.

ABOVE. Diagram of Tesla's AC motor, which he patented in 1888. His early research was instrumental in developing this, his most significant invention.

OPPOSITE. Nikola Tesla in his early 20s, around the time he left his studies in Prague and moved to Budapest, where he would suffer his great nervous breakdown.

BUDAPEST

Without funds to continue his studies, Tesla then moved to Budapest, where he planned to seek work with a family friend who was starting a new telephone exchange. There, perhaps due to the strain of having failed in school and disappointed his parents, he suffered what appears to have been a nervous breakdown, which was associated with extreme acuity of the senses.

I could hear the ticking of a watch with three rooms between me and the time-piece. ... The sun's rays, when periodically intercepted, would cause blows of such force on my brain that they would stun me. ... My pulse varied from a few to 260 beats per minute, and all the tissues of my body [were afflicted] with twitchings and tremors. ... A renowned physician ... pronounced my malady unique and incurable.

Tesla never dared imagine that "so hopeless a physical wreck" could recover, yet a friend of his convinced him that he needed to get outdoors and exercise, and it was there that he both regained his health and achieved his most important breakthrough in the development of his world-renowned invention – the rotating alternating current (AC) magnet. Tesla later claimed that the insight flashed into his mind as he was reciting these words of Goethe's *Faust*:

The glow retreats; done is the day of toil;
It yonder hastes, new fields of life exploring;
Ah, that no wing can lift me from the soil,
Upon its track to follow, follow soaring!

Tesla's key insight was to use two electrical circuits instead of one, and to stagger their position so that they were 90 degrees out of phase with each other. In this way, the motor would be kept rotating continually. In his autobiography, Tesla describes his experience of the discovery in this way:

For a while I gave myself up entirely to the intense enjoyment of picturing machines and devising new forms. It was a mental state of happiness about as complete as I have ever known in life. Ideas came in an uninterrupted stream and the only difficulty was to hold them fast. ... I delighted in imagining the motors constantly running, for in this way they presented to the mind's eye a fascinating sight.

PARIS

Having achieved a breakthrough in the development of alternating current, Tesla longed to share his innovations with the most famous inventor of the day, a man whose name was inextricably attached to electricity and the many new inventions it could power: Thomas Edison. Edison's company had recently formed in Paris and planned to construct generating stations, manufacture lightbulbs, and build lighting stations. The company's work would get under way in the French capital, helping to solidify its reputation as "the city of lights", but eventually spread throughout Europe.

On the advice of a friend, Tesla moved to Paris to take a job with Edison's company, where he would solve problems and help to improve the efficiency of power stations in France and Germany. While on the road, Tesla continued to develop and refine plans for his AC motor, and soon built his first prototype. Tesla became more and more convinced that no one who learned of his design – least of all Thomas Edison – could possibly fail to perceive its superiority.

OPPOSITE. Components from Tesla's AC motors and other devices, as exhibited at the 1893 Columbian Exposition in Chicago.

ABOVE. Artist's depiction of Tesla conducting a demonstration in Paris, where he moved in 1882 to work with Edison.

THOMAS EDISON

Born in Ohio in 1847, the seventh and last of his parents' children, Thomas Alva Edison grew up in Michigan. His family was not well off, and Edison's formal schooling amounted to no more than a few months. Perhaps due to childhood ear infections, Edison developed profound hearing loss, which caused his teachers to label him "slow". As a boy, Edison sold sweets and newspapers on the railways. He eventually published his own newspaper for a time before becoming a telegraph operator.

After marrying and starting a family, in 1870 Edison moved to New Jersey, where he invented the phonograph. News of the invention spread quickly across the country, and soon Edison was in Washington, D.C. demonstrating its ability to record and play back sound to US government officials. Capitalizing on his fame, Edison built a laboratory at Menlo Park, New Jersey, and soon he was widely known as "the wizard of Menlo Park".

Menlo Park was a remarkable operation, the first of its kind explicitly created for the purpose of technological innovation. Over the years, the work at Menlo Park (often regarded as the first industrial research laboratory) helped Edison acquire 1,093 US patients, a record that stood for nearly 100 years. In fact, many of these inventions were developed at least in part by Edison's employees, whose efforts he supervised and directed. Edison developed a reputation as a rigid taskmaster who expected results.

Most of Edison's inventions related to utilities, including the generation and transmission of power,

ABOVE. Photograph of Thomas Edison as a boy.

OPPOSITE. Photograph by Matthew Brady of Thomas Edison with his new invention, the phonograph, 1870s.

but additional devices he developed or improved were Alexander Graham Bell's microphone, the electric light bulb, and the motion picture camera. With the help of financier J.P. Morgan, Edison founded the Edison Electric Light Company in 1878, and his first large commercial installation of electric lighting occurred aboard a steamship, the *Columbia*.

Edison then focused his attention on developing power distribution stations, opening facilities in both London and New York in 1882. Edison's stations powered street lamps and the homes of dozens of nearby residential customers. Yet his DC-based model of power generation meant that electricity could be transmitted over only short distances, which would require the installation of a power station in every neighbourhood.

Describing his method of innovation, Edison famously said, "Invention is ninety-nine percent perspiration and one percent inspiration." Considering that, at his death, he left behind 3,500 notebooks packed with observations, new ideas and sketches, this declaration is not difficult to believe. Edison often worked over 100 hours per week, and his wife set up a cot in his library so he could take brief naps, rather than falling asleep at his workbench.

One of the secrets of his genius seems to have been his refusal to regard any setback as an irrevocable failure. He treated each disappointment as a learning opportunity and persevered in circumstances that might lead others to give up. In the search for the best filament for his electric light bulb, for example, Edison tested thousands of different materials before producing a carbonized cotton filament that could burn for 15 hours. He subsequently lengthened filament life much further.

Edison excelled at business as well as invention and was always looking for ways to capitalize financially on his innovations. One of his most successful business innovations was the company that eventually became General Electric. One of 12 companies listed in the original Dow Jones Industrial Average, it is the only one that remained listed through 2018. Of course, Edison's inventions spawned whole industries, such as electric utilities, recorded music and motion pictures.

Though Tesla initially regarded Edison as a hero, both the spirit of rivalry that would arise between them and Tesla's ultimate appraisal of his own superiority as an inventor are captured in his words from a 1931 *New York Times* article:

> *If Edison had a needle to find in a haystack, he would proceed at once with the diligence of the bee to examine straw after straw until he found the object of his search. ... I was a sorry witness of such doings, knowing that a little theory and calculation would have saved him ninety percent of his labor.*
>
> (New York Times, *19 October 1931)*

ABOVE. Photograph of three of the most successful entrepreneurs of the early twentieth century: (L to R) Henry Ford, Thomas Edison, and Harvey Firestone.

AMERICA

Tesla believed that he had served Edison's company in Paris admirably by solving a number of design problems with the company's dynamos and he deserved to be rewarded. Yet to his great disappointment, no such reward was forthcoming. So Tesla decided to pack his bags and go to the "Land of Golden Promise", the United States. From the start, his journey appeared ill-fated. When he showed up at the train station, both his money and tickets were gone. Running alongside the train, Tesla's resolve wavered, but his dexterity won out in the end, and soon he was bound for America.

The year he landed in America was 1884, and he went almost immediately to meet Edison, who was facing difficulties with a steam ship, the SS *Oregon*, for which he had supplied the lighting. Tesla went to the ship, worked all night to get the generators operating again, and left at 5 am. On his way home, on the street he encountered Edison, who assumed Tesla had been out carousing. When Tesla told him the ship was now operational, Edison told one of his assistants, "This is a damn good man."

Later, when Edison learned of Tesla's habit of working from 10:30 am to 5:00 am the next morning, Edison told him, "I have had many hard-working assistants, but you take the cake." The Serbian was given wide latitude and improved the design of 24 different machines for the company. Tesla reports that he was promised a reward of $50,000 on the completion of this task, but when he announced his success, he was told that it had been a practical joke. Angry, he abruptly resigned.

The Edison–Tesla relationship had been strained from the start, thanks to dramatic differences in personality. Tesla was a well-educated person who spoke many languages and savoured classic works of literature. Edison was an immensely practical man with little use for "eggheads". In his heyday, Tesla was the toast of society, while the hard-of-hearing and introverted Edison was not much of a conversation partner. And Tesla was highly fastidious, while Edison could be neglectful of his own hygiene.

TOP. The Edison Machine Works in lower Manhattan as it appeared in the early 1880s.

ABOVE. The SS *Oregon*, a passenger/cargo ship on which Edison's company had installed an electrical lighting system that was repaired in a single night in 1884, by a crew directed by Tesla.

Spkg Telegraph

July 18 1877
Chas Batchelor
James Adams

X is a rubber membrane connected to the central diaphragm and the edge being near or between the lips in the act of spkg it gets a vibration which is communicated to the central diaphragm & thus in its turn set the outer diaphragm again vibrating hence the [illegible] consonants are reinforced & made to set the diaphragm in motion we just tried an experiment similar to this one

Just tried experiment with a diaphragm having an embossing point & held against paraffin paper moving rapidly the spkg vibrations are indented nicely & theres no doubt that I shall be able to store up & reproduce automatically at any future time the human voice perfectly

183

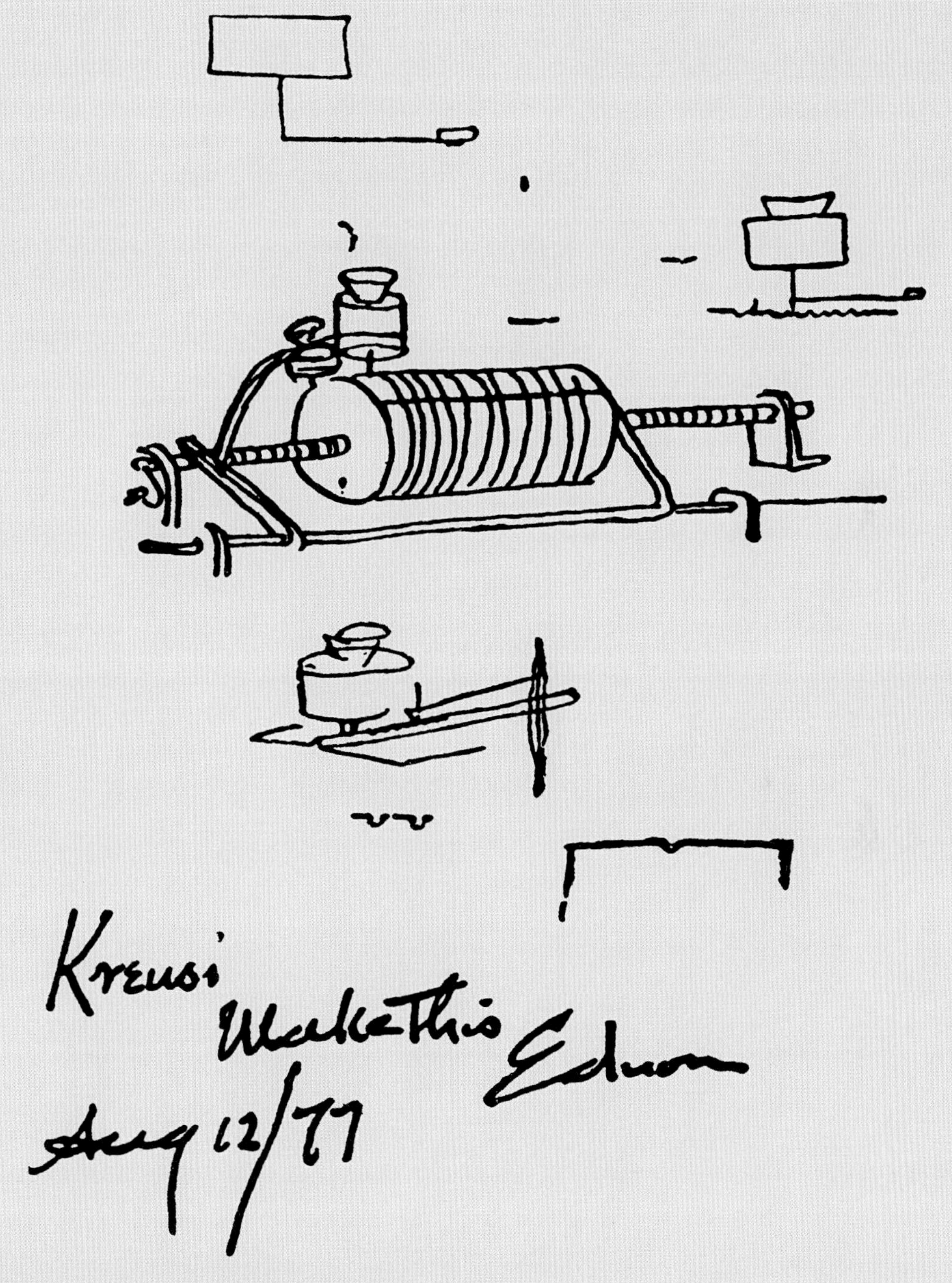

OPPOSITE. Edison's first notes on the phonograph, as recorded on one of the pages of his extensive notebooks.

ABOVE. Edison's sketch of the first phonograph, with instructions for his modeller, Kruesi, as drafted in 1877.

THE DAWN OF AC

Soon after he left Edison's company, Tesla was approached by a group of investors who offered to back him in starting his own company. Tesla naturally assumed that they coveted his AC motor, but he was soon disappointed to learn that they were seeking an improved arc lamp. So he worked on an improved municipal arc lighting system, which was installed in the town of Rahway, New Jersey, lighting streets and factories.

Eventually, Tesla became so frustrated with his backers' lack of interest in his AC motor that he was forced out of the firm that bore his own name, the Tesla Electric Light and Manufacturing Company. Though he had "perfected" the arc system, he was left with "no other possession than a beautifully engraved certificate of stock of hypothetical value". Bankrupt, for a time Tesla was forced to go to work as a ditch-digger. He felt his great education and ambitions were only mocking him.

In 1887, the foreman of Tesla's crew became convinced of his abilities as an engineer and connected him with a telegraph engineer, who in turn reached out to a prominent lawyer for help in establishing another business to pursue Tesla's dreams. To Tesla's mortification, however, the lawyer, Charles Peck, showed no interest in AC power. Prospects looked bleak until Tesla's knowledge of culture and history came to his aid.

He asked Peck if he remembered the story of the Egg of Columbus. According to legend, the explorer Christopher Columbus had gained the support of Queen Isabella of Spain for his proposed expedition by setting and solving a riddle. He challenged the members of Isabella's court to balance an egg on its end. None could. Then Columbus slightly cracked the end of the egg shell and was able to make it stand unsupported. Duly impressed, Isabella pawned her jewels to finance the expedition.

Tesla assured Peck that he could solve the same riddle, only better. When Peck took the bait, Tesla had a metal egg made, constructed a circular apparatus with polyphase currents running around its perimeter, and then went back to Peck to demonstrate. At first, the egg began to spin and then to wobble. But as Tesla increased the current further, the wobbling ceased, and the egg began spinning gracefully on its end. The result? The birth of the Tesla Electric Company.

As Tesla envisioned it, his new company would design and produce not only individual components of electrical equipment, such as generators, motors and transformers, but whole AC systems that were designed to function together. The first AC system

OPPOSITE. William Hogarth's "Columbus Breaking the Egg" (1752).

BELOW. Tesla's Egg of Columbus apparatus, as illustrated in the March, 1919 issue of the *Electrical Experimenter* magazine.

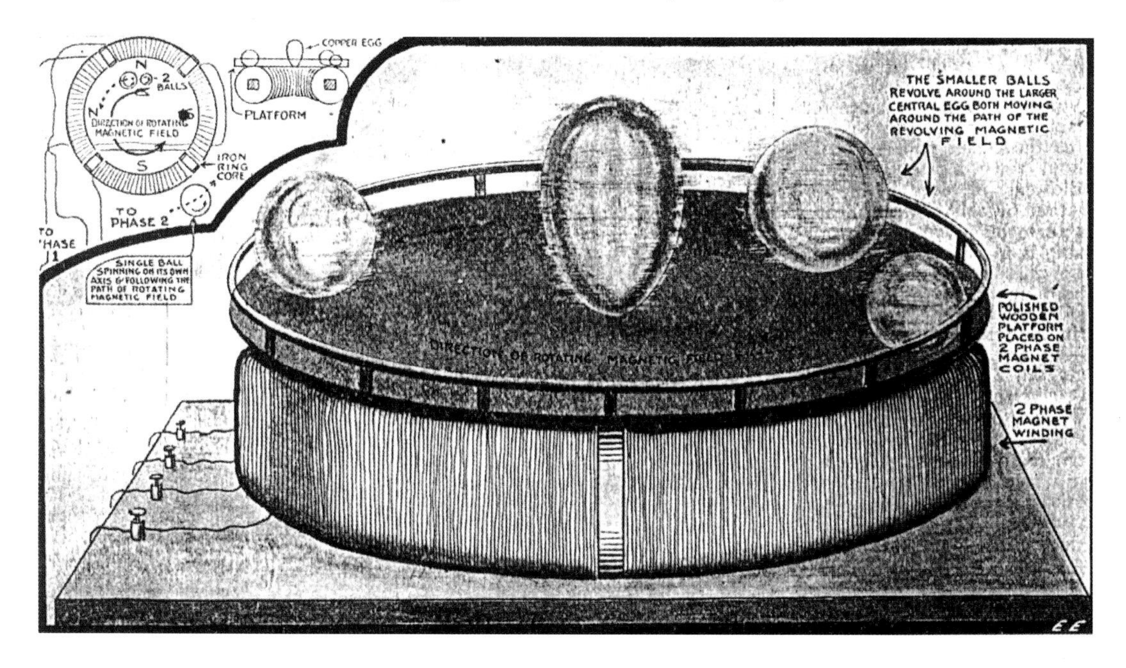

(No Model.) 2 Sheets—Sheet 1.

N. TESLA.

ELECTRIC ARC LAMP.

No. 335,786. Patented Feb. 9, 1886.

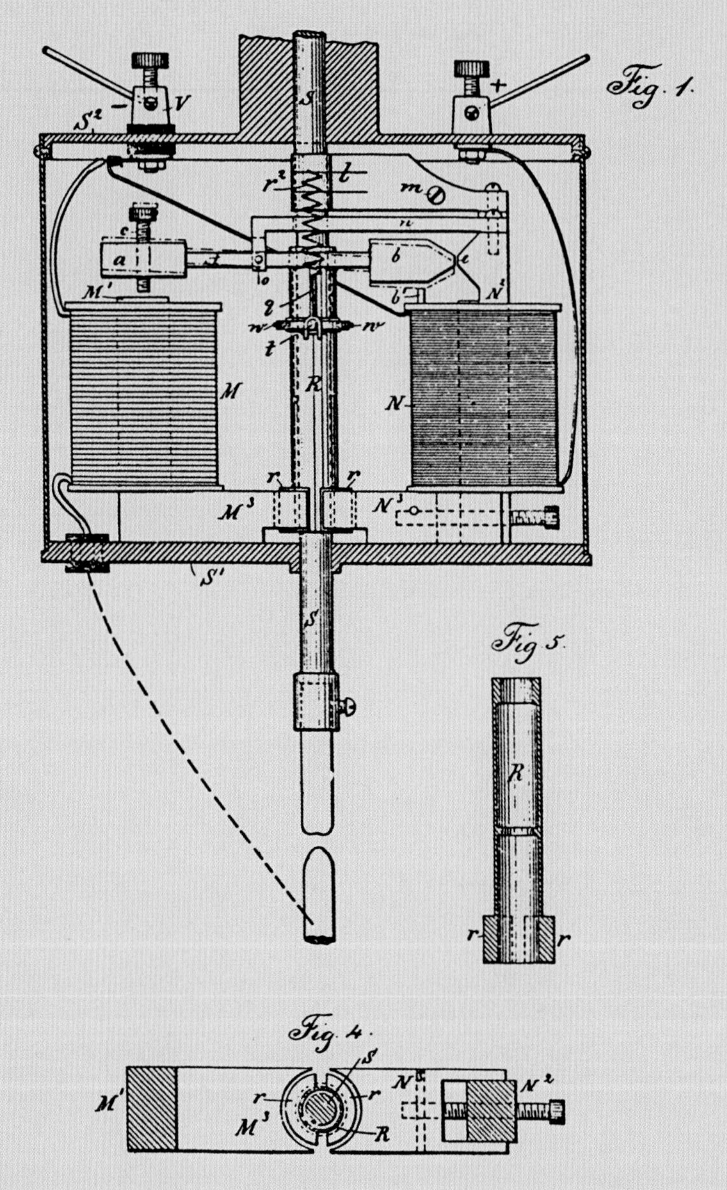

Witnesses:

J. Staib

Chas H. Smith

Inventor

Nikola Tesla

per Lemuel W. Serrell

atty.

OPPOSITE. A diagram accompanying Tesla's patent application for an electric arc lamp.

RIGHT. An example of one of Tesla's original induction motors, as exhibited at a meeting of the American Institute of Electrical Engineers at Columbia University in 1888.

had been installed by the Westinghouse company in Buffalo, NY the year before, and just one year later he had over 30 plants in operation. Tesla's company had some catching up to do.

In 1888, fortune smiled on Tesla. A Cornell University Professor who established one of the nation's first courses in electrical engineering had been extolling the virtues of Tesla's induction motor. He was particularly impressed that it functioned at high efficiency, losing relatively little energy to heat, and it also broke down infrequently. Thanks to this support, Tesla was invited to address the American Institute of Electrical Engineers (AIEE).

His lecture was entitled "A New System of Alternate Current Motors and Transformers", and it proved to be a smash hit. After a few introductory remarks, Tesla began his address with these words:

> *The subject which I now have the pleasure of bringing to your notice is a novel system of electric distribution and transmission of power by means of alternate currents, affording peculiar advantages, particularly in the way of motors, which I am confident will at once establish the superior adaptability of these currents to the transmission of power and will show that many results heretofore unattainable can be reached by their use; results which are very much desired in the practical operation of such systems and which cannot be accomplished by means of continuous [DC] currents.*

Commenting years later on the importance of Tesla's address, the electrical engineer (and AIEE Vice-President) Bernard Behrend wrote in the minutes of a May 1917 AIEE meeting:

> *Not since the appearance of Faraday's "Experimental Researches in Electricity" has a great experimental truth been voiced so simply and so clearly. ... Tesla left nothing to be done by those who followed him.*

Representatives of the Westinghouse company were on hand; they wasted no time in reaching out to the man who seemed to have perfected an AC power system that would make possible the transmission and efficient utilization of electrical power over distances reaching perhaps hundreds of miles.

GEORGE WESTINGHOUSE

It is difficult to exaggerate the influence of railways over the course of world history. Prior to the nineteenth century, large quantities of goods and people generally needed to be transported over water – oceans, lakes or rivers. With the coming of the "iron horse", longer land distances became traversable in shorter periods of time. Many historians regard the 1869 completion of the transcontinental railroad in Utah as one of the greatest engineering feats of the nineteenth century.

The history of rail transport extends back to Greece in the sixth century BC. Centuries later, wooden rail systems along which cars were pulled using rope and a treadwheel were in use in sixteenth century Europe. An early breakthrough occurred in the 1700s, when metal replaced wood in rails. Then, in the late eighteenth century, James Watt introduced steam-powered locomotives, which could be used to transport much heavier cargoes. Largely unsolved, however, was the problem of stopping the train.

It is said that the seeds for the railroad air brake were sown when a young George Westinghouse witnessed a collision between two trains. The engineers saw each other from a distance and did what they could to avoid the crash, cutting power to the locomotive and using their whistles to notify brakemen, who would jump from car to car, applying the brake in each one. Despite their best efforts, however, the trains collided, resulting in injuries and loss of life.

Relying on brakemen to apply the brakes in each car was not only dangerous but ineffective, and

trains often stopped too soon or too late for a station. Westinghouse, then in his early 20s, felt certain that there must be a better way. His "*Eureka!*" moment came to him one day when he was reading about the use of compressed air drills to tunnel through a mountain. His idea, which he patented in 1869, was to use hoses carrying compressed air to connect and apply the brakes in each car.

Later, hoping to improve the safety of the system even further, Westinghouse turned this principle upside down. Instead of relying on air pressure to apply the brakes, he constructed a system in which air pressure would keep them off. In this way, if the supply of compressed air to the train or a car were ever interrupted, the brakes would be applied passively. First installed in 1872, this new system ensured that loose cars would stop.

The worldwide introduction of Westinghouse's system made stopping trains both safer and more precise. Providing assurance that trains could be stopped, it also meant that longer ones could travel faster along the tracks. Westinghouse made additional contributions to rail transportation, including the invention of electrical signals that prevented two trains from occupying the same segment of track, which further lowered the risk of collision.

Westinghouse was born in New York in 1846, and his gifts for both invention and business were recognized early. After enlisting for military service

ABOVE. Photograph depicting the celebration of the completion of the first transcontinental railroad in Promontory Point, Utah on 10 May, 1869.

OPPOSITE. Photograph of George Westinghouse from the early twentieth century.

in the Civil War at age 15 and later dropping out of college, he created his first invention, the rotary steam engine, at age 19. He was 22 when he first introduced the air brake system, and his air brake and rail signal systems made him one of the best-known inventors and wealthiest men of his day.

As had been the case with his work on the air brake, Westinghouse became interested in power distribution somewhat by accident. In 1883 at his home in Pittsburgh, drillers tapped into a small vein of natural gas in his back yard. Later, in the middle of the night, Westinghouse was awakened by an explosion. When he inspected the site, he found all the drilling equipment gone, destroyed by a geyser that at first emitted mud but later began issuing a jet of pure natural gas.

This situation persisted for a week, until Westinghouse devised a stopcock that brought the well under control. This led to his first patent for a system of conveying and utilizing gas under pressure, which was followed by patents for improved techniques for digging gas wells, meters for measuring the amount of gas used, and regulators for controlling the mixture of gas and air. Not surprisingly, he also developed a shut-off valve to stop the flow of gas whenever pressure fell too low, suggesting a leak.

Over the course of his life, until his death in 1914 at age 68, Westinghouse acquired 361 patents, founded 60 companies, and amassed a large fortune. Unlike fellow titans of industry Andrew Carnegie and Henry Ford, Westinghouse maintained good relations with his workforce, and was the first industrialist to shorten the work week from 6 to 5 ½ days. Among his contemporaries, perhaps only Ford rivals him in terms of success as both an inventor and a businessman, though he did not hold nearly so many patents.

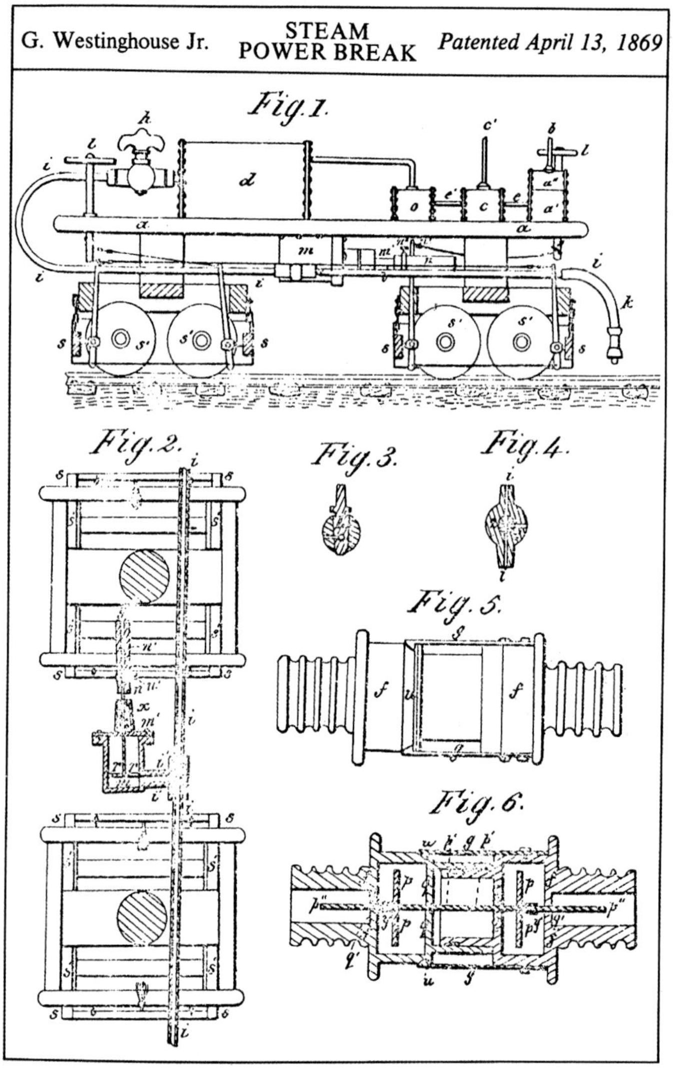

ABOVE. The 1869 patent for Westinghouse's railway air brake.

OPPOSITE ABOVE. Diagram of Westinghouse's quick-action, triple valve, fail-safe air brake, as introduced in 1887.

OPPOSITE BELOW. Artist's depiction of an early assembly line producing air brakes at the Westinghouse foundry in Pittsburgh in 1890.

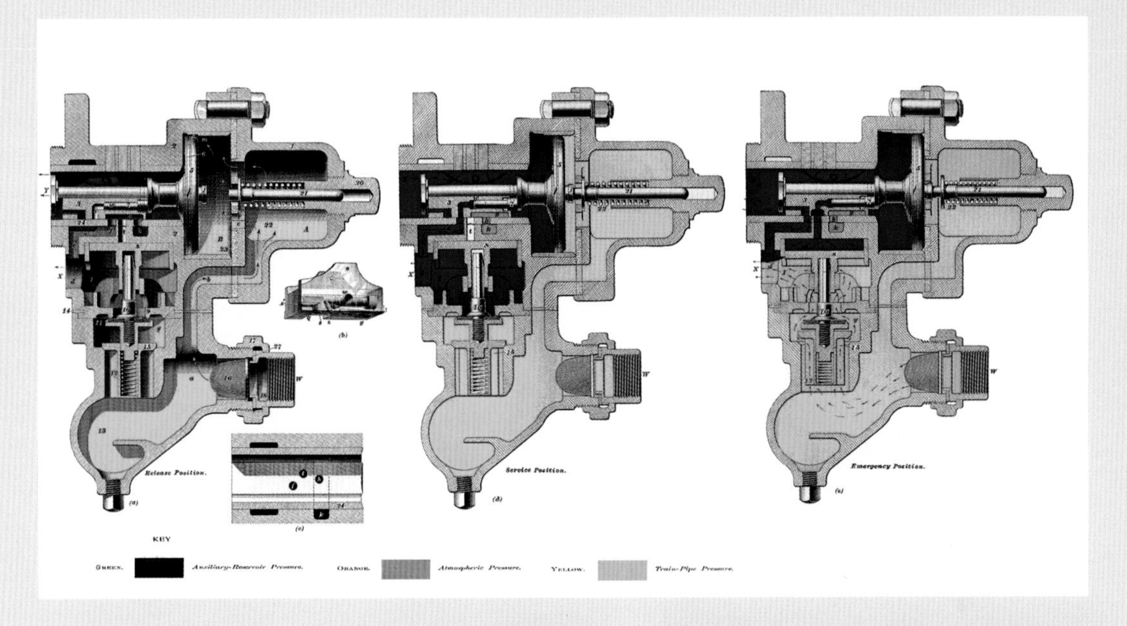
Release Position.
Service Position.
Emergency Position.
KEY
GREEN.
Auxiliary-Reservoir Pressure.
ORANGE.
Atmospheric Pressure.
YELLOW.
Train-Pipe Pressure.

TAKING AC TO SCALE

Westinghouse was convinced that the future of electricity was AC, and that Nikola Tesla's patents would prove indispensable to whatever firm acquired them. So he instructed his representatives to obtain the patents, and they eventually did so, though not before agreeing to pay Tesla the exorbitant sum of $2.50 for every horsepower generated by machines based on his design. Tesla journeyed to Pittsburgh to finalize the agreement with Westinghouse.

Westinghouse favourably impressed Tesla, who in a 21 March 1914 article for *Electrical World*, described him in these terms:

> *A powerful frame, well proportioned, with every joint in working order, an eye as clear as a crystal, a quick and springy step – he presented a rare example of health and strength. ... An athlete in ordinary life, he was transformed into a giant when confronted with difficulties which seemed insurmountable. He enjoyed the struggle and never lost confidence. When others would give up in despair, he triumphed. ... He gave to the world a number of valuable inventions and improvements, created new industries, advanced the mechanical and electrical arts and improved in many ways the conditions of modern life.*

As part of the agreement Westinghouse and Tesla hammered out, Tesla agreed to move to Pittsburgh. He regretted leaving behind the life of New York's upper social strata, to which he had just been introduced, but residing in Pittsburgh would allow him to play an active role in developing the AC system. The agreement also specified that Westinghouse would defend Tesla's patents against

ABOVE. One of the 5,000-horsepower 2-phase generators installed by Westinghouse at Niagara Falls.

OPPOSITE ABOVE. One of the Tesla induction motors manufactured by the Westinghouse Company.

OPPOSITE BELOW. Westinghouse's laboratory where Tesla developed equipment for use in AC systems.

challenges, and he was soon busy doing so, as other innovators tried to establish their own priority.

There was, however, a problem with Tesla's design. The existing AC power plants that Westinghouse and others had installed around the country operated at a relatively high frequency of 133 Hertz, while Tesla's designs were based on a lower frequency of 60 Hertz. While there were strong arguments for Tesla's approach, rendering the two systems compatible was an expensive proposition, and Westinghouse was becoming overextended financially.

Tesla, frustrated that Westinghouse was not developing the AC system, eventually agreed to forego the $2.50 per horsepower royalty. He did so not because he was indifferent to the vast fortune it seemed destined to generate, but because he believed so strongly in the potential of his system to change the world. His vision of the benefit it would provide mankind was so palpable to him that he could not fail to agree to remove such an obstacle to its realization.

(No Model.)

N. TESLA.

ALTERNATING CURRENT MOTOR.

No. 433,701. Patented Aug. 5, 1890.

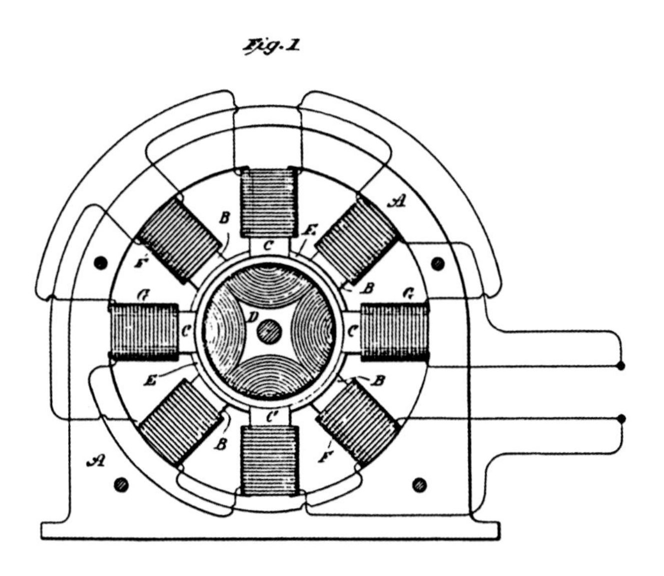

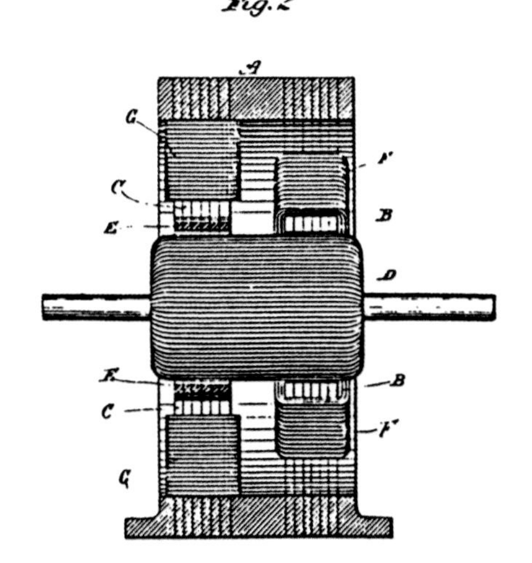

Witnesses:
Raphael Netter
Ernest Hopkinson

Inventor
Nikola Tesla
by
Duncan, Curtis & Page
Attorneys.

T 240

Family Name	Given Name or Names
TESLA	NIKOLA

Title and Location of Court
COMMON PLEAS COURT, NEW YORK COUNTY.

Date of Naturalization	~~Volume or~~ Bundle No.	Page No.	Copy of Record No.
JULY 30-1891	701	—	35

Address of Naturalized Person
HOTEL GERLACH W. 27 ST.-N.Y.C.

Occupation	Birth Date or Age	Former Nationality
CIVIL ENGINEER	—	AUSTRIAN

Port of Arrival in the United States	Date of Arrival
—	JUNE 1884

Names, Addresses and Occupations of Witnesses To Naturalization	
1	RICHARD F. FEIST RAHWAY, N.J.
2	

OPPOSITE. Patent for another of Tesla's many AC motors, 1890.

ABOVE. Tesla's US naturalization record, which he regarded as at least equal in importance to any of his inventions.

THE WAR OF THE CURRENTS

In retrospect, it seems obvious that the Westinghouse-backed AC system of electrification would triumph over the Edison-backed DC (direct current) system, but at the time the issue was hotly contested. One of the principal problems with DC was the fact that power could be transmitted over relatively short distances. By contrast, using step-up and step-down transformers, Westinghouse could step up power for long-distance transmission and then step it back down for local use.

Unable to achieve victory among scientists and engineers, Edison decided to wage war in the court of public opinion. When workers on high-voltage AC power lines suffered injuries and deaths, Edison did his best to fan the flames of public alarm, seeking to paint AC as an inherently dangerous form of power transmission. He sought to portray the proponents of AC as greedy businessmen, prepared to sacrifice lives for the sake of profits.

To further dramatize the danger of AC, Edison helped to fund research on the use of AC to electrocute animals. Because there were so many stray animals on the streets of large cities, electrocution soon became a widely employed means of putting them down. Soon people were exploring the idea that similar means could provide a quicker and more humane means of executing condemned criminals. Soon Edison was referring to the technique as "Westinghousing" the subjects.

When the electric chair was invented for executing criminals, Edison pulled out all the stops to ensure that the device would be powered by a Westinghouse generator. In 1889 a new term was introduced into the world's lexicon: electrocution. If Edison had his way, AC would soon become known

OPPOSITE. Press photograph of the 1903 execution of Topsy the circus elephant.

BELOW. An electric chair used at Auburn State Prison in New York, site of William Kemmler's execution in 1890.

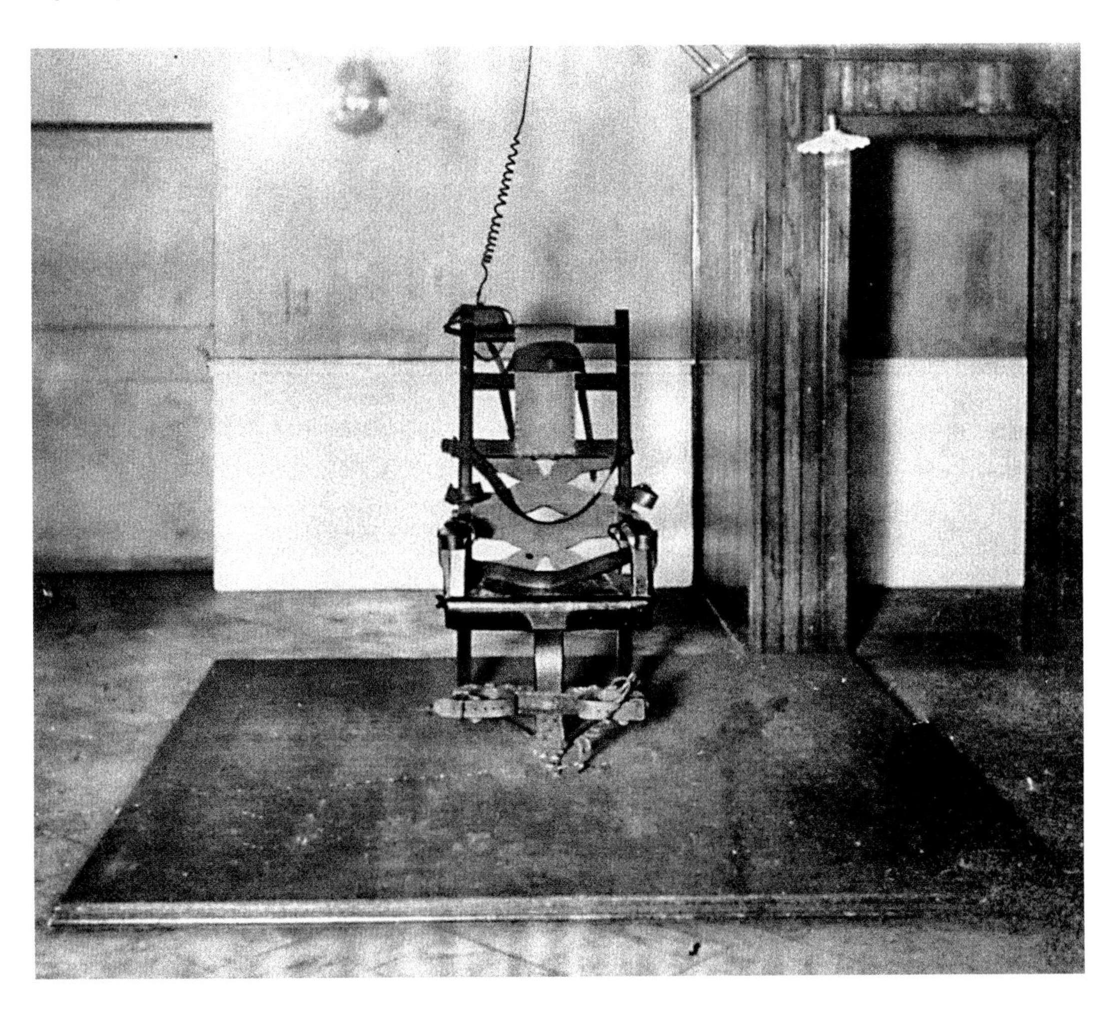

as the electrocutioner's current. The first execution by electricity would take place in August of 1890.

The condemned criminal was a man called William Kemmler, who had been convicted of murdering his common-law wife with a hatchet. When Westinghouse learned what was happening, he hired a lawyer to represent the condemned man, in an effort to put a stop to matters. However, the appeal failed. Proceeding with the execution, current at 1,000 volts was applied for 17 seconds. Kemmler was thought to be dead, but it soon became apparent that he was still breathing. A second application at 2,000 volts was ordered. The execution required 8 minutes.

A similar spectacle was staged in 1903. Topsy, an Asian circus elephant that had been involved in several injuries including the death of a spectator, was scheduled to be put down. Her original owners had planned to charge admission to her hanging, but the Society for the Prevention of Cruelty to Animals put a stop to it. So an electrocution was staged instead. On hand was a crew from the Edison Motion Picture Company, which filmed the event.

Today electrocution is still an accepted method of execution in some US states, some of which allow the condemned individual to choose between this method and lethal injection. Other states, such as Nebraska, have deemed it a "cruel and unusual" form of punishment and abandoned it. Reflecting on its use in Kemmler's case, Westinghouse later said, "They would have done better using an axe", and a reporter who witnessed the event described it as "an awful spectacle, far worse than a hanging".

RIGHT. Demonstration of the world's first electric chair, at Auburn Prison, Auburn, New York, which was first used to execute William Kemmler on 6 August, 1890.

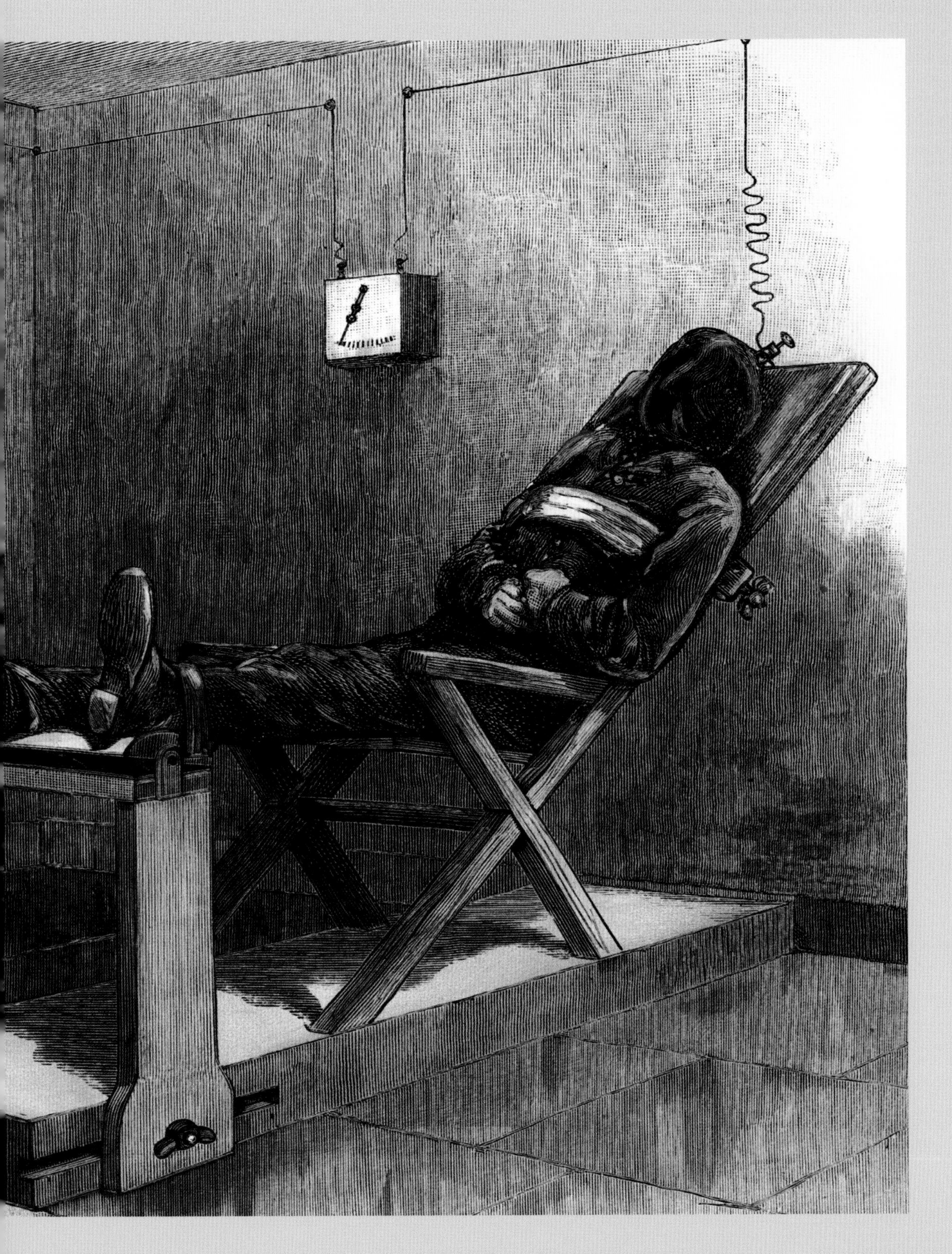

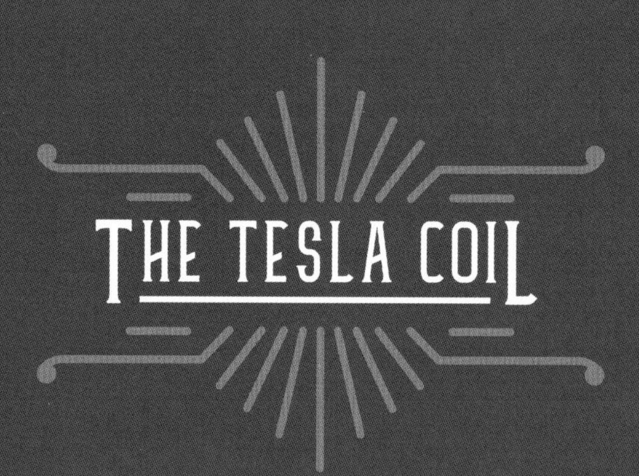

THE TESLA COIL

Tesla dreamed of the wireless transmission of electrical power, and he made a leap in this direction with the invention of the Tesla coil, which he patented in April of 1891. Tesla first publicly demonstrated the device at a meeting of the American Institute of Electrical Engineers, which took place at Columbia College in New York in May of that year. The device features an induction coil that can produce high-frequency alternating currents capable of producing lightning-like electrical discharges.

A Tesla coil consists of two principal parts: a primary coil and a secondary coil. The two coils, which represent independent electrical circuits, are separated by a spark gap. High voltages are obtained with the use of a transformer. When the device is turned on, power accumulates in the primary coil's capacitor. Once it reaches a certain point, current flows at high voltage across the spark gap into the secondary coil, until enough energy builds up to transfer it back again.

If the two coils resonate, these transfers of energy gradually build – just as tiny pushes can make a playground swing go higher and higher – with transfers occurring hundreds of times per second. Because the secondary coil is connected to a capacitor, generally located at the top of the device, when the charge in the secondary coil becomes sufficiently high, it can be released in the form of lightning-like discharges, which Tesla used to astonish spectators.

Tesla coils have energies ranging from several thousand to several million volts, if the coils are large enough. The current is typically in a relatively low frequency range, at the radio end of the electromagnetic spectrum (see p.64). Moreover, although the voltages (energies) can be huge, the current (amount of electricity) is usually small, which helps to explain how Tesla could allow it to flow through the human body without serious health effects.

The lightning-like discharges escaping into the air around a Tesla coil are streamer arcs, formed when

ABOVE. Members of the Tesla Coil Builders Association with their largest coil, nicknamed "Nemesis." The association was founded by Harry Goldman of Queensbury, NY, in 1982.

OPPOSITE. Photograph showing streamers emitting from one of Tesla's wireless transmitters. This is a resonant wireless transmitting circuit, with a secondary flat spiral coil connected to an antenna.

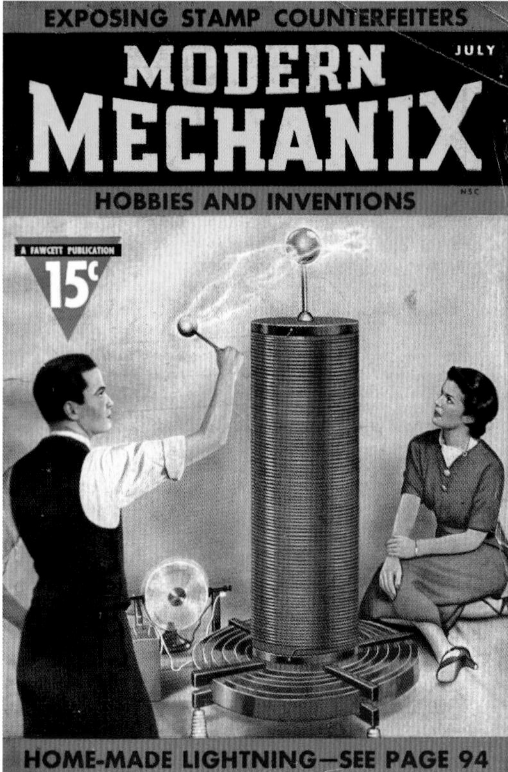

the strength of the electric field surpasses the capacity of air to serve as an insulator. At this point, air itself degrades and becomes a conductor. As the voltage increases, these electrical discharges can travel further out into the air, but a point is reached at which air can conduct them no further. Coils immersed in oil, however, can produce even longer discharges.

The electrical discharges themselves exhibit a branching pattern, and their length and colour vary depending on the voltage and current. High-voltage, low-energy discharges produce patterns that are relatively short and purple or blue in colour, while high-voltage, high-energy discharges tend to be both longer and white in colour. Such discharges produce ozone and nitrogen oxides detectable as distinct odours in the vicinity of the discharges.

Tesla and his contemporaries used the ability to transmit electrical energy wirelessly in a variety of ways. They investigated lighting, phosphorescence (glow-in-the-dark materials), and X-ray production. Such circuits were used in wireless telegraphs for several decades. Tesla himself used the device in part to establish and build his reputation as a wizard who could manipulate fundamental forces of nature in amazing ways.

ABOVE LEFT. A Tesla coil presented by Lord Kelvin to fellow scientists in 1897.

ABOVE RIGHT. Cover of the July 1937 issue of *Modern Mechanix* magazine, depicting Tesla's vision in the popular imagination.

OPPOSITE. 1899 photograph of Tesla wirelessly illuminating a fluorescent light bulb in the vicinity of a Tesla coil, whose electromagnetic field ionizes the gas in the bulb, producing light.

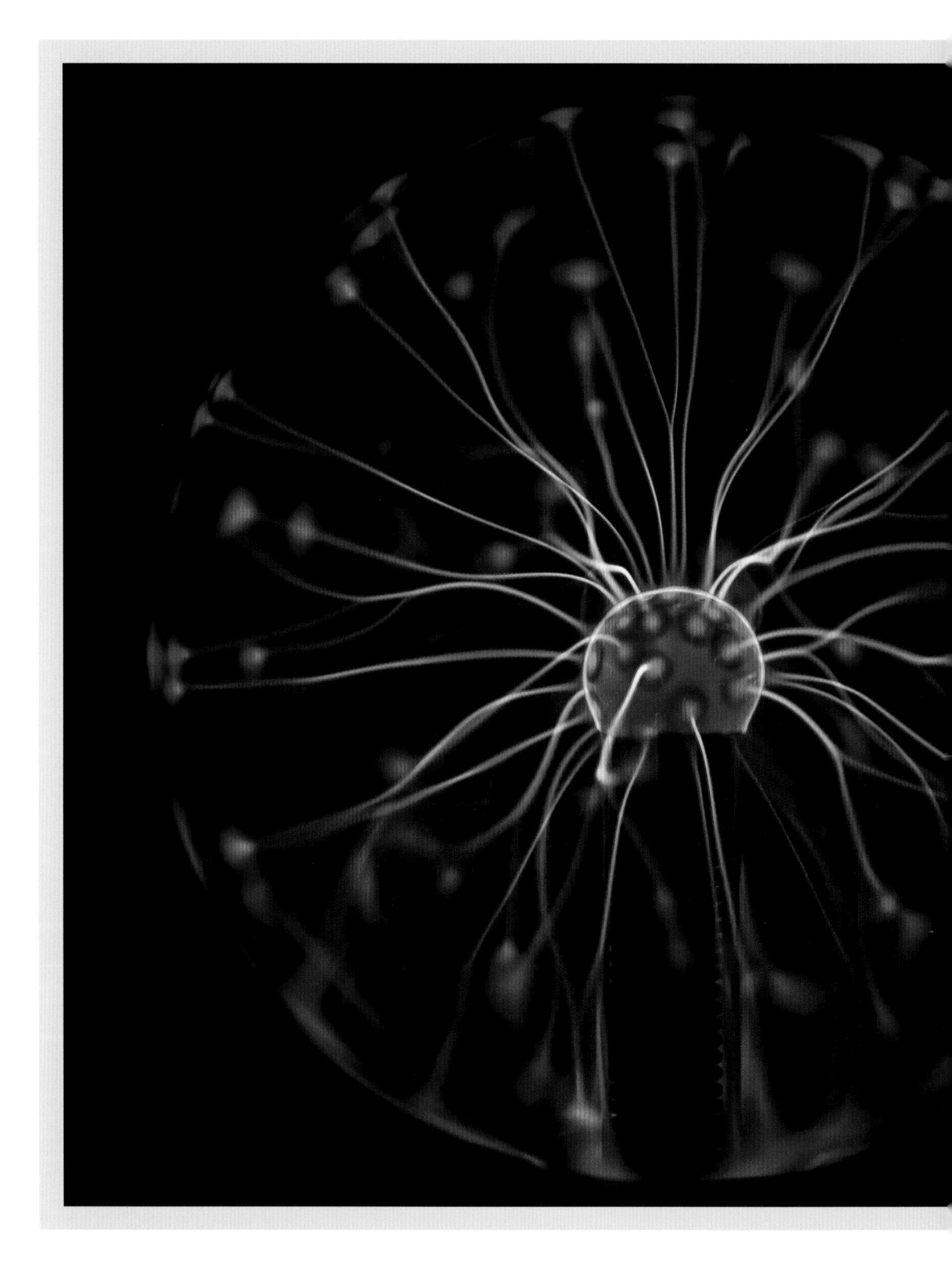

In addition to the Tesla coil's relatively low currents, another reason Tesla could allow such discharges to flow through his body was the fact that the human nervous system is insensitive to electrical currents at high frequencies. By comparison, a 60-cycles-per-second current from a wall outlet can prove quite painful. Yet the fact that Tesla coil discharges can be painless does not necessarily prove that they are completely harmless.

When such discharges strike skin, they can still cause burns. This is one reason why Tesla would employ bars of metal, metallic thimbles on his fingers, or fluorescent light bulbs in his demonstrations. This allows the current entering the body to do so over a sufficiently large surface area that no burns are produced. The current can, however, produce a sense of warmth, which led to its use in diathermy, a therapy that was thought to promote healing.

A modified hand-held Tesla coil can still be used to look for leaks in devices that require a vacuum, such as a vacuum tube. Because air at low pressure conducts electricity better than air at atmospheric pressure, the discharges of the Tesla coil will pass through and illuminate even the tiniest pin-hole. Today, however, Tesla coils tend to be used primarily in educational displays in museums and science fairs, where they serve as popular do-it-yourself science projects.

STRANGE MANIFESTATIONS

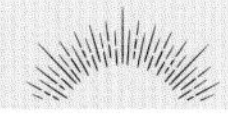

As the following excerpt from one of Tesla's addresses indicates, public amazement at the first demonstration of the Tesla coil was matched by Tesla's own astonishment at the forces he was wielding.

> *Of all the forms of nature's immeasurable, all-pervading energy, which ever and ever [are] changing and moving; like a soul animates the inert universe, electricity and magnetism are perhaps the most fascinating. The effects of gravitation, of heat and light we observe daily, and soon we get accustomed to them, and soon they lose for us the character of the marvellous and wonderful; but electricity and magnetism, with their singular relationship, with their seemingly dual character, unique among the forces in nature, with their phenomena of attractions, repulsions and rotations, strange manifestations of mysterious agents; stimulate and excite the mind to thought and research. What is electricity, and what is magnetism? These questions have been asked again and again. The most able intellects have ceaselessly wrestled with the problem; still the question has not as yet been fully answered.*
>
> *(Lecture delivered before the American Institute of Electrical Engineers, Columbia College, NY, 20 May 1891)*

The answers to these questions would continue to occupy Tesla to the end of his long life.

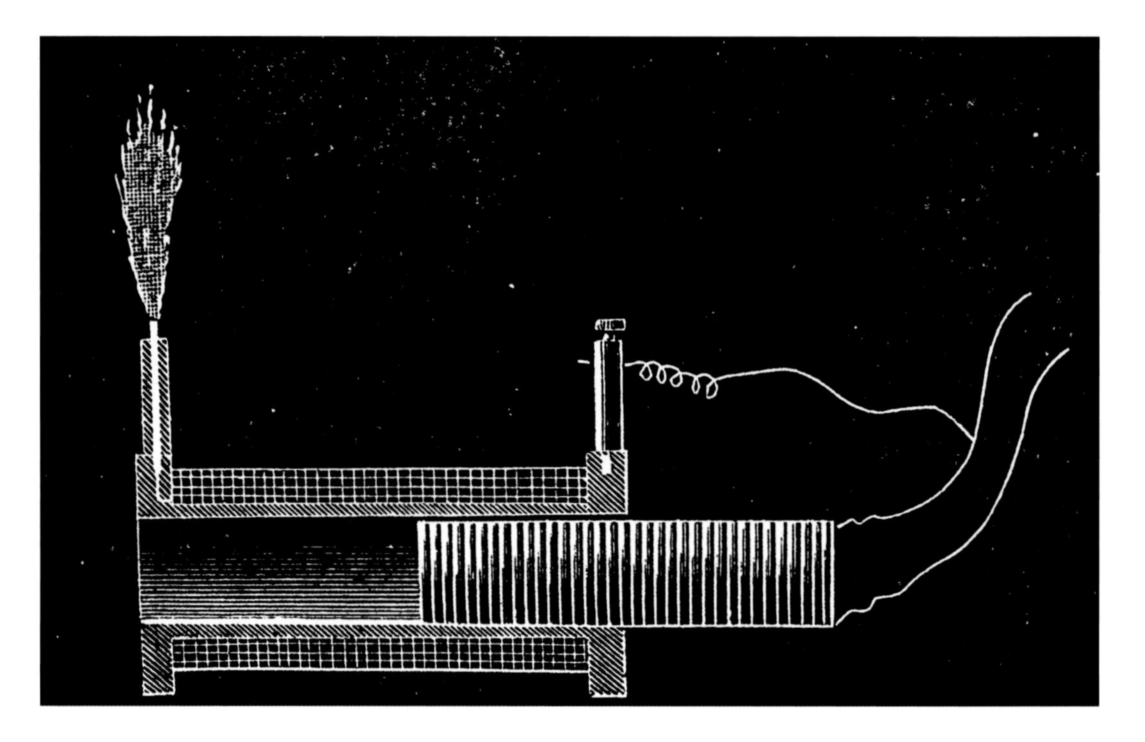

ABOVE. Artist's depiction of St Elmo's Fire as produced by one of Tesla's AC experiments.

OPPOSITE. Diagram from one of Tesla's patent applications for an electrical condenser and coil, 1897.

(No Model.)

N. TESLA.

MANUFACTURE OF ELECTRICAL CONDENSERS, COILS, &c.

No. 577,671. Patented Feb. 23, 1897.

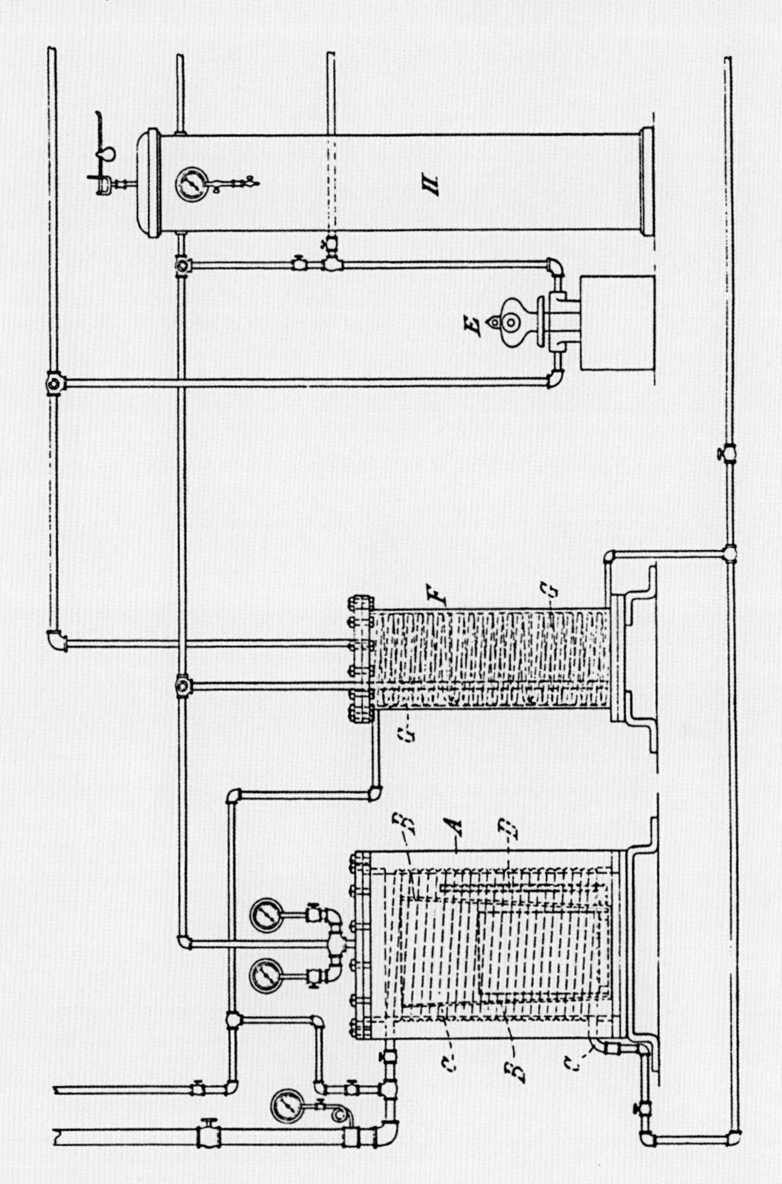

WITNESSES

Edwin B. Hopkinson,

Benjamin Miller,

INVENTOR

Nikola Tesla

BY

Ker. Curtis & Page

ATTORNEYS

1893 WORLD'S COLUMBIAN EXPOSITION

The 1893 World's Columbian Exposition, also known as the Chicago World's Fair, celebrated the 400th anniversary of Christopher Columbus's arrival in the New World in 1492. The fair grounds covered nearly 283 hectares (700 acres) on the south side of Chicago, and between the fair's opening on 1 May and conclusion on 30 October it attracted 27 million visitors. Because of the alabaster-hued materials used to cover the fair's 14 great buildings, it became known as the White City.

It was clear to the two principal combatants in the War of the Currents (see p.42) that winning the contract to supply electricity to the fair would provide a once-in-a-lifetime showcase for the victor. Edison General Electric put in a bid of $554,000 to light the fair with DC current, but Westinghouse came in with a bid to do the same with AC for only $399,000. Thanks to Westinghouse, Tesla's polyphase system would be on display for the whole world to see.

There was just one problem: Edison, perhaps offended at the loss of the contract, would not allow Westinghouse to use his patented incandescent light bulbs. Under immense pressure, Westinghouse came up with a design that replaced Edison's sealed glass globe with a bulb having a ground glass stopper in one end. While these bulbs provided more than satisfactory illumination, they burned out quickly, requiring battalions of workers to keep replacing them throughout the fair.

As a result of this enormous effort, the Chicago Fair did more to light the night than any man-made event in human history. The switch to power on the fair's 130,000 incandescent bulbs for the first time was thrown by US President Grover Cleveland. Powering the fair were 12 one-thousand-horsepower Tesla AC polyphase generators, and one of the exhibition buildings was Electricity Hall, in front of which stood a statue of Benjamin Franklin.

OPPOSITE. Photograph of the interior of Electricity Hall at the 1893 Chicago World's Fair.

ABOVE, TOP. Photograph of the 1893 World's Columbian Exposition in Chicago, colloquially referred to as the White City.

ABOVE. The Westinghouse AC switchboard at the 1893 Columbian Exposition in Chicago.

Inside Electricity Hall were exhibits from the major corporations of the day, including Edison General Electric and Westinghouse. Edison's exhibit featured his kinetoscope, an early version of a motion picture device, and his phonograph, which could record and play back voices and music, as well as a variety of electrically powered household appliances. In the middle of General Electric's exhibit stood a 25-metre (82-foot)-tall "Tower of Light" made from 30,000 pieces of cut glass.

Ever the showman, Tesla relished the opportunity to display his inventions. Housed in the Westinghouse display, the Tesla exhibit included examples of his AC motors and generators, vacuum tubes illuminated by the wireless transmission of power, and his famous Egg of Columbus (see p.31). The exhibit also contained a device that produced lightning-like discharges and the matching thunderous reports, which could be heard throughout the hall.

When Tesla's visit to the fair was announced, it was claimed that he would "send a current of 100,000 volts through his own body without injury", a feat made more remarkable by the fact that only several thousand volts were necessary for electrocutions. In doing so, Tesla was relying on the since-discredited hypothesis of the "skin effect" – the mistaken notion that high-frequency currents tend to remain on the surface of the body, thereby averting internal damage.

In August, Tesla made a presentation at the fair to the International Electrical Congress, whose participants included some of the foremost electrical scientists and engineers in the world. Public excitement at the event ran high, and many spectators crowded the doorways but were denied admittance. To the fascination of those present, Tesla demonstrated his new mechanical oscillators, which could harmonize motors and electrical generators and even permit the wireless transmission of information.

According to one engineer in attendance,

> *Mr. Tesla was very much elated over his latest achievements and said that he hoped that in the hands of practical as well as scientific men, the devices described by him would yield important results. He laid special stress on the facility now afforded for investigating the effect of mechanical vibration in all directions.*

Tesla was quoted as saying that he regarded his investigations into oscillatory phenomena as his "greatest work", and that he hoped "to continue his investigations until he could present to the world a system of mechanical and electrical vibration of commercial utility". No one knew it at the time, not even Tesla, but he was describing the foundations of what would later become known as wireless communication, including radio and television.

PREVIOUS PAGES. Night-time photograph of the Agricultural Hall at the 1893 World's Fair in Chicago, the site of Tesla's presentation to the International Electrical College on 25 August.

OPPOSITE. Tesla transmitting electricity wirelessly through his body in 1899.

BELOW. The wireless lighting tubes exhibited by Tesla at the 1893 Columbian Exposition in Chicago.

MAGIC SHOW

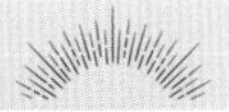

In addition to promoting scientific and technological progress, Tesla also provided a great spectacle that appealed to the aesthetic sense of those who attended the fair. One fair-goer described Tesla's exhibit in these terms:

> *Large glass plates backed with tin foil, on which are outlined, in paper, various figures, are used, and on them the play of the electric spark produced effects that are dazzling and extremely beautiful. A voltage of 30,000 is used up to the condensers, and after it leaves them it is estimated that the current has a power of two million volts. Mr. Tesla also shows a number of other interesting experiments, some of which are so marvelous as to be almost beyond description.*

Because of such displays, Tesla developed a rather distinctive public persona that incorporated disparate elements. He was garnering fame as an engineer, inventor, and to some degree scientist, but he also acquired a reputation resembling that of a conjurer or wizard, someone who enjoyed amazing audiences with feats of wonder. The tension between science and magic, technology and legerdemain, left many wondering what precisely to make of Nikola Tesla.

NIAGARA FALLS

Straddling the border between Canada and the United States are three of the most famous waterfalls in the world, known collectively as Niagara Falls. Extending from Lake Erie into Lake Ontario and standing 50 metres (165 feet) tall, Niagara has a flow rate of 170,000 cubic metres (6 million cubic feet) of water per minute. It was created about 10,000 years ago by the Wisconsin glaciation, and as erosion of rock continues over the next 50,000 years, the falls will erode back to Lake Erie and cease to exist.

Numerous daredevils have attempted to survive the Falls in a barrel or to traverse portions of it by tightrope, and in recent years, visitors to Niagara Falls have numbered about 30 million annually, most of whom are drawn by its majestic beauty. But beginning in the eighteenth century, innovators began regarding the Niagara as a source of power, first for mills but later for electrical power.

The first hydroelectric station was built in 1881 in Appleton, Wisconsin. Today hydroelectric plants produce approximately 17 per cent of the world's electricity, most notably at the Three Gorges Dam in China, which ranks as the world's largest power-production facility of any kind. Of course, one of the greatest benefits of hydroelectric power is the fact that it is non-polluting and completely renewable, which helps to explain why it accounts for about 70 per cent of the world's renewable electricity production.

Tesla claimed to have dreamed first of travelling to America and harnessing Niagara's power in his youth, and he got his chance in the early 1890s. In 1890, an

ABOVE. Photograph of William Thomson, Lord Kelvin (1824–1907), one of the most famous physicists of his day, who decided that Westinghouse offered the best proposal to harness the power of Niagara.

OPPOSITE. Niagara Falls as viewed from the Canadian side.

A MONUMENT OF ENLIGHTENMENT

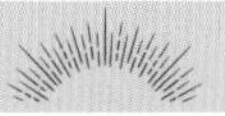

Musing on the power harnessed by his polyphase system, Tesla said in his 12 January 1897 speech at the opening ceremony for the hydroelectric power station at Niagara Falls:

> *We have many a monument of past ages; we have the palaces and pyramids, the temples of the Greeks and the cathedrals of Christendom. In them is exemplified the power of men, the greatness of nations, the love of art and religious devotion. But the monument at Niagara has something of its own, more in accord with our present thoughts and tendencies. It is a monument worthy of our scientific age, a true monument of enlightenment and of peace. It signifies the subjugation of natural forces to the service of man, the discontinuance of barbarous methods, the relieving of millions from want and suffering.*

BELOW. Interior of the Niagara Falls power station.

international commission of experts led by eminent Scottish physicist and engineer Lord Kelvin began to study how best to generate electricity there. They issued a request for proposals and received 19 from teams backed by some of the world's wealthiest men. The commission rejected them all before announcing another contest in 1893 that garnered proposals from Edison and Westinghouse.

The balance was tipped in favour of Westinghouse in part because Lord Kelvin visited the World Columbian Exposition in Chicago in 1893, which was powered by Tesla's polyphase system. It took more than two years to complete the project, but generators capable of producing 50,000 horsepower were eventually installed, and in November of 1896 a switch was thrown that transmitted power from the Falls to Buffalo, New York, 20 miles away.

Ultimately, the energy powering hydroelectric plants comes from the sun. The sun's radiant energy heats water in oceans and lakes, causing it to evaporate. When that water falls to the earth in the form of rain, some of it collects at relatively high points on the earth's surface. As it flows from higher to lower points, such as from Lake Erie into Lake Ontario, its kinetic energy can be tapped to turn wheels to grind grain or turbines to turn electrical generators.

To tap into the vast quantities of energy contained in such large quantities of water falling such long distances, the water was channelled into tunnels lined by turbines that could convert the water's kinetic energy into electrical energy. Eventually, power was transmitted into New York City, where it powered trams, subways, and the lights of Broadway. It is estimated that within 25 years, 25 per cent of US electrical power was generated by hydroelectric plants.

The Niagara project did much to establish Tesla's popular reputation. Though he had not designed the project, Tesla laid the groundwork for AC and the polyphase system as the best means of harnessing the Falls. Journalists competed with one another in singing his praises, leading one to label Niagara "the unrivalled engineering achievement of the nineteenth century". The *New York Times* described Tesla's contribution in these terms:

> *Even now the world is more apt to think of him as a producer of weird experimental effects than as a practical and useful inventor. Not so the scientific public or the business men. By the latter classes Tesla is properly appreciated, honored, perhaps even envied. For he has given to the world a complete solution of the problem which has taxed the brains and occupied the time of the greatest electro-scientists of the last two decades – namely, the successful adaption of electrical power transmitted over long distances.* (New York Times, *July 1895)*

The contract to construct the lines that transmitted the power from Niagara to Buffalo was awarded to Edison's company – a highly ironic choice, since Edison's preference for DC represented one of the principal reasons that he lost the contract to build the Niagara hydroelectric plant in the first place. Edison may have been stubborn but he was educable, though, and soon after the Niagara project was completed he began to embrace AC power.

BELOW. The Tesla statue in Niagara Falls, Ontario, depicting the inventor atop one of his AC motors. There is also a statue of Tesla on the American side of Niagara Falls.

X-RAYS

Wilhelm Roentgen, who won the first Nobel Prize in Physics for his discovery of the X-ray, may have been the first scientist to investigate and describe the properties of X-rays in a systematic fashion, but he was certainly not the first to notice the peculiar effects of this "new" form of electromagnetic radiation. When Roentgen announced his discovery at the end of 1895, Tesla immediately realized that the German physicist was describing the same "radiant energy of an invisible kind" that he himself had noticed.

X-rays represent a portion of the electromagnetic spectrum, which stretches from very long-wavelength, low-frequency radio waves to very short-wavelength, high-frequency gamma waves, with visible light in between. X-rays are near the high-frequency end of the spectrum. Beyond their most familiar use in medical radiology, X-rays have played a crucial role in elucidating the structure of molecules through X-ray crystallography, and the structure of the universe through X-ray astronomy.

By 1894, Tesla had been experimenting with Crookes tubes and had invented his own form of vacuum tube, both of which are now known to produce X-rays. Tesla believed that streams of what we now call electrons produced the new form of radiation when they struck an object, such as the glass wall of the tube he had invented. He hypothesized that this new form of radiation was composed of particles, which we now call photons.

It is impossible to know where Tesla's investigations

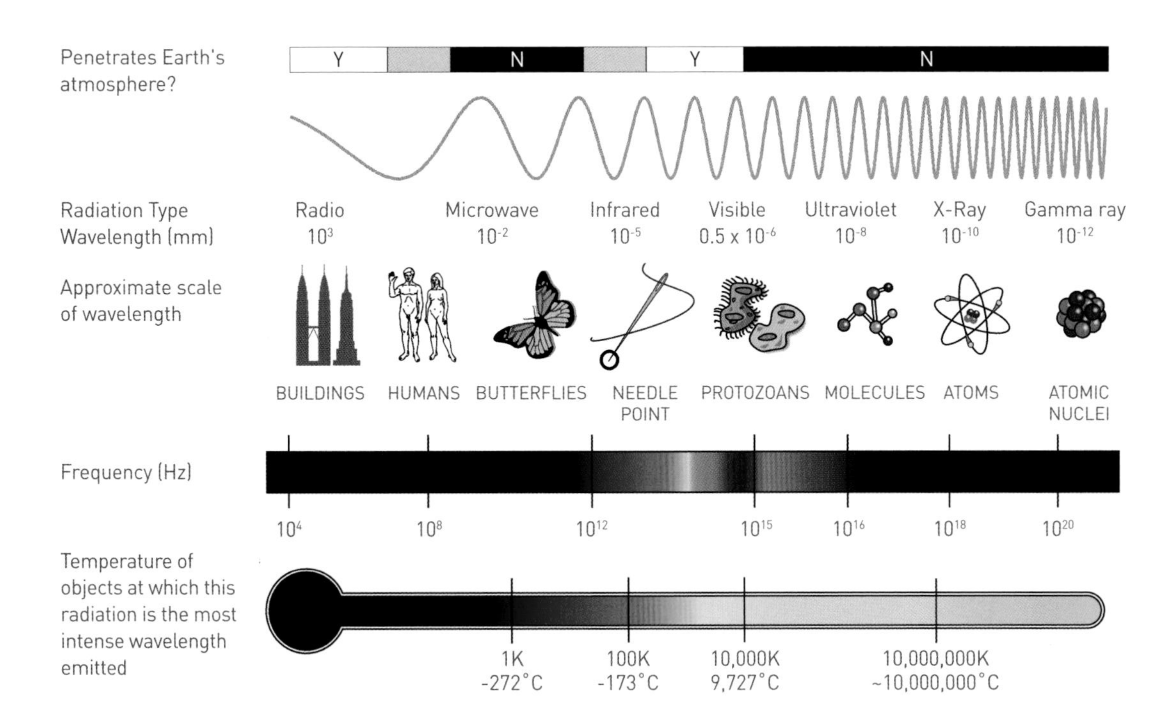

of the new form of radiation would have taken him, but we do know that they were brought to an abrupt halt when Tesla's laboratory was consumed by fire in March 1895. The announcement of Roentgen's discovery rekindled Tesla's interest in the subject, however, and he continued his experiments with X-rays for several years thereafter, using different vacuum tubes of his own design.

Tesla did more than generate X-rays; he also produced some of the first X-ray images. Mark Twain (see p.70) had been fascinated with Tesla's work since the early 1890s, and he was a frequent visitor to Tesla's laboratory. One day Tesla tried to

ABOVE. The electromagnetic spectrum.

RIGHT. Nineteenth-century artist's depiction of how an X-ray image of the hand could be produced using Roentgen's X-Ray Machine.

OPPOSITE. Wilhelm Conrad Roentgen (1845–1923), the German physicist who in 1901 won the first Nobel Prize for the discovery of X-rays, and who corresponded with Tesla about the discovery.

produce a photographic image of Twain using one of his vacuum tubes. However, the images showed not Twain but metallic components of the camera. This failed attempt to image Twain in fact represents one of the earliest X-ray images.

Within months of Roentgen's announcement, Tesla was producing high-quality human images. He called them "shadowgraphs", an entirely apt expression because X-ray images are composed of the shadows cast by different anatomical structures as the X-ray beam passes through the body. Some of Tesla's images show human anatomy, including the rib cage and a foot as seen through a shoe.

Tesla sent some of his human X-ray images to Roentgen, who responded with a note thanking him (pictured overleaf), which read:

> *Dear Sir,*
> *You have surprised me tremendously with the beautiful photographs of wonderful discharges and I tell you thank you very much for that. If only I knew how you made such things! With the expression of special respect, I remain your devoted*
> *W. C. Roentgen*

Tesla recognized potential roles for X-rays in health and disease. For example, he recognized that X-ray images could be used to determine the position of foreign objects such as bullets within the human body. He also expressed the view that X-rays could help detect diseases of the lung. Unlike Roentgen, however, who set aside everything to focus entirely on X-rays, Tesla's attention was shared with many other pursuits.

Tesla also realized that X-ray exposure could be harmful to health – producing, for example, inflammation of the skin not unlike sunburn. Though he incorrectly attributed these effects to gases produced by X-ray generation rather than the radiation itself, he also recognized other harmful effects, such as irritation of the eyes. In addition, he correctly argued that these adverse effects could be reduced by increasing the distance from the X-ray source or decreasing the time of exposure.

A WELL-GROOMED MAN

Tesla's work on X-rays revealed not only interior anatomy but also something of his external appearance. A February 1896 *New York Herald* article occasioned by Tesla's work on X-rays also contains one of the best descriptions of Tesla himself:

> *Mr. Tesla is tall and slender, with very black, thick hair. He wears a very small mustache over a small mouth. All of his features are delicate. His cheeks are rather hollow, but they have the flush of health. His cheekbones are conspicuous. His manner is cordial, and as attractive in its way as his appearance, for he is not the careless begrimed machinist as he issues from his shop, but a "well groomed" man, whose clothing is stylish and well fitting, and whose hands and face are as pink and clean as a baby's. Like most inventors, he is a late worker. The forces of both mind and body, he thinks, operate better at night than in the daytime. So he is often in his shop until two or three o'clock in the morning. He seldom comes down before noon. Occasionally he is attracted by "society" and he was one of the star features at a recent [reception] at which Sarah Bernhardt also shone. But his work absorbs most of his time, and he has little taste for anything else.*

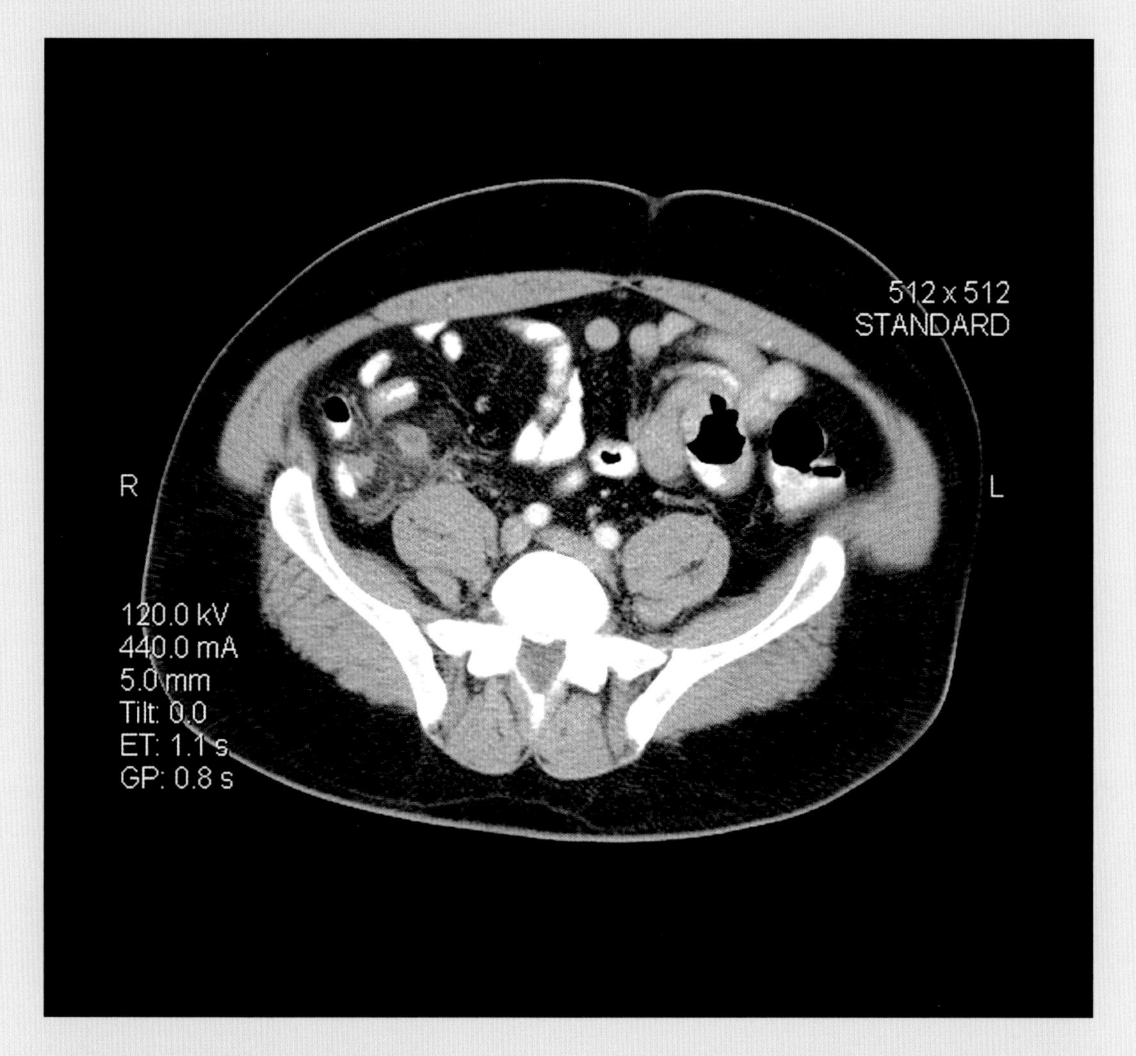

Tesla's concerns were shared by Thomas Edison, though only much later. Aghast at the severe radiation necrosis experienced by his glassblower and laboratory assistant, Clarence Dally, who eventually died from a radiation-induced cancer, Edison (who experienced damage to his own vision from X-ray exposure) later wrote:

> *I did not want to know anything more about X-rays. In the hands of experienced operators, they are a valuable adjunct to surgery... But they are dangerous, deadly, in the hands of [the] inexperienced, or even in the hands of a man who is using them continuously for experiment.* (New York World, *3 August 1903)*

Despite the hazards of X-rays and other medical imaging techniques based on them, such as CT scanning, today their advantages hold sway in the popular imagination. Common medical diagnoses such as bleeding within the brain, pneumonia and appendicitis are routinely based on the use of these techniques, and the introduction of X-ray imaging is frequently ranked among the most important innovations in the history of medicine.

ABOVE. CT scan of the abdomen in a patient with acute appendicitis. The dilated and inflamed appendix, which is located just to the viewer's left of the 17.1 mm measurement, normally measures fewer than 7 mm in diameter.

13 München 20 Juli 1901

MUZEJ NIKOLE TESLE

Herrn Nicola Tesla in New York.

Sehr geehrter Herr!

Sie haben mich ungemein überrascht mit den schönen Photographien von den wunderbaren Entladungen, und ich sage Ihnen vielen Dank dafür.

Wenn ich nur wüsste, wie Sie solche Sachen machen!

Mit dem Ausdruck besonderer Hochachtung verbleibe ich

Ihr ergebener

W.C. Röntgen

ABOVE. Roentgen's 1901 letter to Tesla.

RIGHT. 1896 X-ray image of a human foot, often thought to be Tesla's own

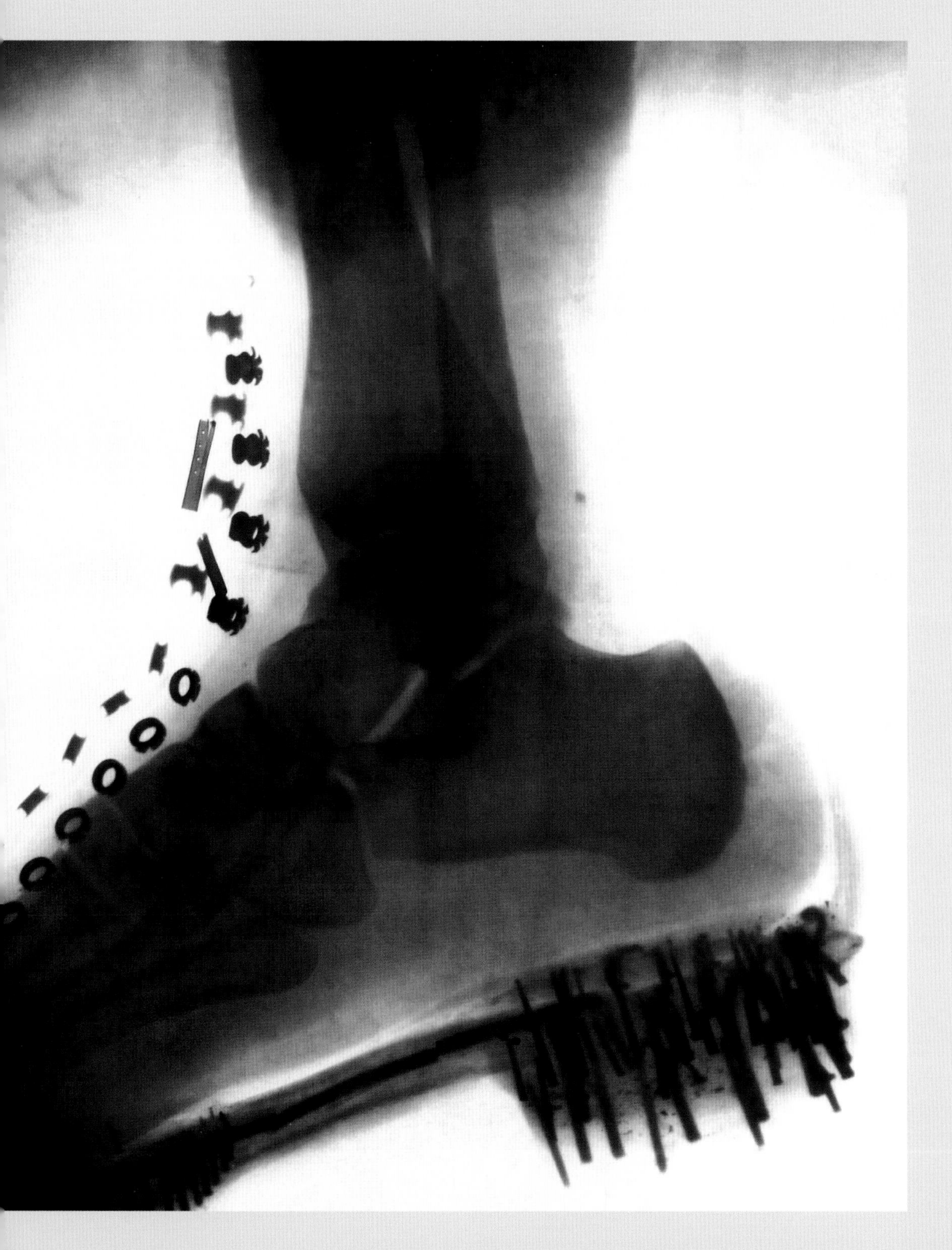

Born in the small town of Florida, Missouri in 1835, Samuel Langhorne Clemens had become famous worldwide as the writer Mark Twain by the time he died in Connecticut in 1910. He was raised in the Missouri town of Hannibal on the Mississippi River, which would provide the setting for his two most famous novels – *The Adventures of Tom Sawyer* and *The Adventures of Huckleberry Finn* – and the tale of his life is at least as remarkable as any of the stories he told.

Twain's father died when he was 11, and just a year later he began working in the printing business, which gave him his start as a writer of humorous sketches. At 18 he began moving from job to job in large American cities, reading widely in public libraries at night. At 21 he began his apprenticeship as a river boat pilot, earning his licence two years later. At this point he began making frequent use of the pilot's phrase "mark twain", meaning that the river is deep enough to navigate safely.

Twain persuaded his brother, Henry, to join him on the river, but the young man was killed at age 20 in an explosion. With the outbreak of the American Civil War in 1861, Twain headed west, working as a miner and then as a newspaper writer. It was in Nevada in 1863 that he first used his pen name, Mark Twain. He broke through as a writer in 1865 with the story "The Celebrated Jumping Frog of Calaveras County". Soon Twain became a world traveller, filing dispatches from Hawaii, Europe, and the Middle East.

It was while travelling that he first saw a photograph of his future wife, Olivia, a young New York woman of privilege who initially rebuffed his advances but eventually succumbed to his raffish charm. After the two were married, their son Langdon died at 19 months, but three daughters survived to adulthood. It was while enjoying an idyllic family life in Hartford, Connecticut that Twain wrote many of his best-known novels.

Though a man of letters, Twain loved technology. His 1889 novel *A Connecticut Yankee in King Arthur's Court* features a time traveller who uses his knowledge of science to introduce the Arthurian Age to modern technologies. He held three patents, one of which, a self-pasting scrapbook, sold well. Despite making vast sums of money, Twain frequently found himself in dire financial straits, largely thanks to his investments in new technologies, most notably a failed typesetting machine.

Financially strapped, Twain and family closed their house in Hartford in 1891 and relocated to Europe,

OPPOSITE. The young Samuel Clemens, typesetter, holding a printer's composing stick with the letters SAM.

although he returned to the States multiple times. It was during these visits that he began spending time in Tesla's New York laboratory. It is said that during one of these visits Tesla invited the chronically constipated Twain to stand atop one of his mechanical oscillators, which soon relieved Twain of his condition.

TOOLS FOR PEACE

Twain and Tesla were both visionaries, men who dreamed of creating a better world. In 1897, while in Austria, Twain learned of Tesla's work on new technologies that could have military applications. He wrote to Tesla to encourage him to think of such innovations of tools for peace, not war:

Dear Mr. Tesla,

Have you Austrian and English patents on that destructive terror which you are inventing? – and if so, won't you set a price upon them and commission me to sell them? I know cabinet ministers of both countries. ...

Here in the hotel the other night when some interested men were discussing means to persuade the nations to join with the Czar and disarm, I advised them to seek something more sure than disarmament by perishable paper-contract – invite the great inventors to contrive something against which fleets and armies would be helpless, and thus make war thenceforth impossible. I did not suspect that you were already attending to that, and getting ready to introduce into the earth permanent peace and disarmament in a practical and mandatory way. ...

Sincerely yours,

Mark Twain

In 1909, Twain foretold his own death, saying, "I came in with Halley's Comet [in 1835]. It is coming again next year [in 1910], and I expect to go out with it. ... The Almighty has said, no doubt, 'Now there are these two unaccountable freaks; they came in together, they must go out together.'" (*Mark Twain: A Biography*, Albert Bigelow Paine). He died on 21 April 1910, just one day after the comet reached its perihelion – a conjunction between biographical and celestial phenomena recalled by the flash of lightning at Tesla's birth.

That Twain meant a great deal to Tesla is clear from the circumstances of Tesla's own death. In 1943, an ailing Tesla asked a messenger to deliver a message to Samuel Clemens on South Fifth Avenue. When the boy returned with the sealed envelope, reporting that he could not find the address, Tesla claimed to have spoken with him recently. Only later did the boy learn that the name of the street had been changed, that Tesla was referring to Mark Twain, and that Twain had been dead for decades.

OPPOSITE. Mark Twain, his wife Olivia, and their three daughters.

BELOW. 1894 photograph from Tesla's laboratory of author Mark Twain holding an illuminated lamp, as Tesla stands behind him.

RADIO CONTROL

Tesla was a genius as both an innovator and a showman, and his talent for showmanship was never in greater evidence than during a demonstration he conducted at New York's Madison Square Garden in 1898. Tesla arranged for a large pool of water to be built in the centre of the floor, and into this he placed a boat that measured approximately 1.2 metres (4 feet) in length and 1 metre (3 feet) in height. Multiple antennas adorned its deck, and beneath were radio receivers and electric motors.

Tesla and his control box stood at one end of the pool. To the astonishment of the audience, Tesla proceeded to demonstrate his ability to control the boat wirelessly. He could command it to move forwards, backwards, turn to the right or left, stop, and turn on and off the lamps he had mounted on it. Tesla invited audience members to shout out instructions, with which Tesla could make the boat comply immediately.

Tesla's efforts were motivated in part by his patriotism. Having become a naturalized United States citizen in 1891, he had a keen interest in assisting the US military, which in 1898 was in the thick of the Spanish–American War. Sent to guard American interests during the Cuban revolt against Spain, the USS *Maine* had been sunk in Havana Harbour in February of that year. Whether the Spanish had been responsible or not, the US press used the sinking as a rallying cry for war against Spain.

A year later, Tesla touted the military potential of his device in these terms:

... my submarine boat, loaded with its torpedoes, can start out from a protected bay or be dropped over a ship side, make its devious way below the surface, through dangerous channels of mine beds, into protected harbors and attack a fleet at anchor, or go out to sea and circle about, watching for its prey, then dart upon it at a favorable moment, rush up to within a hundred feet if need be, discharge its deadly weapon and return to the hand that sent it. Yet all through these wonderful evolutions it will be under the absolute and instant control of a distant human hand on a far-off headland, or on a warship whose hull is below the horizon and invisible to the enemy. ... I am aware that this sounds almost incredible, and I have

OPPOSITE LEFT. Aerial view of the interior of Tesla's 1898 radio-controlled boat.

OPPOSITE RIGHT. Photograph of the radio-controlled boat Tesla demonstrated at Madison Square Garden in 1898.

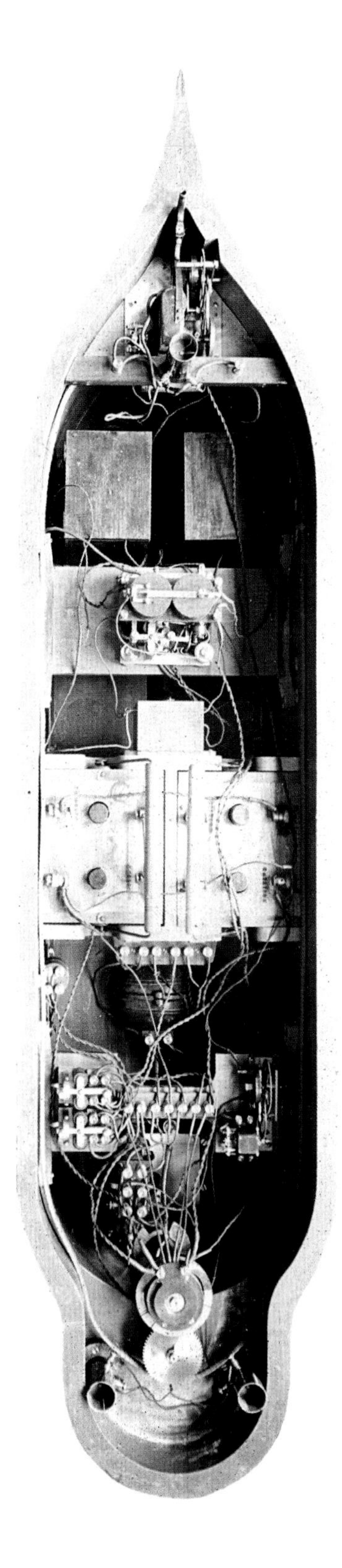

refrained from making this invention public till I had worked out every practical detail of it. (New York Journal, *13 November 1898)*

What Tesla was demonstrating was both an advanced form of wireless communication and one of the early instances of what might be called a radio-controlled robot. The term "robot" would not be introduced until Karel Čapek's 1920 play *R.U.R.* (Rossum's Universal Robots), from a Czech term meaning "forced labour". But what Tesla had invented was a machine capable of carrying out a complex series of actions as prescribed by a human controller, in this case requiring no physical connection.

Tesla referred to his device as a "teleautomaton", a word derived from the Greek root *tele*, meaning "over a distance", and *automaton*, meaning "self-moving". In his view, such devices would reduce the loss of human life in warfare and enable the development of weapons of such overwhelming force that war itself might become a less tenable option. He dreamed, in other words, of a world in which technological innovation rendered war obsolete.

But Tesla's robotic vision was even more radical than this. In a June 1900 article in *The Century Magazine*, Tesla described how he

... conceived the idea of constructing an automaton which would mechanically represent me, and which would respond, as I do myself, but, of course, in a much more primitive manner, to external influences. Such an automaton evidently had to have motive power, organs for locomotion, directive organs, and one or more sensitive organs so adapted as to be excited by external stimuli. The machine would, I reasoned, perform its movements in the manner of a living being, for it would have all the chief mechanical characteristics of elements of the same. ... So this

invention was evolved, and so a new art came into existence, for the which the name "teleautomatics" has been suggested, which means the art of controlling the movements of distant automatons. This principle was evidently applicable to any kind of machine that moves on land or in the water or in the air. In applying it practically for the first time, I selected a boat. ... Evidently the automaton should respond only to an individual call, as a person responds to a name. Such considerations led me to conclude that the sensitive device of the machine should correspond to the ear rather than the eye of a human being, for in this case its actions could be controlled irrespective of intervening obstacles, regardless of its position relative to the distant controlling apparatus, and last, but not least, it would remain deaf and unresponsive, like a faithful servant, to all calls but that of its master. ... By the simple means described the knowledge, experience, judgment – the mind, so to speak – of the distant operator were embodied in that machine, which was thus enabled to move and to perform all its operations with reason and intelligence.

The automatons so far constructed had "borrowed minds," so to speak, as each merely formed part of the distant operator who conveyed to it his intelligent orders, but this art is only the beginning. I purpose to show that, however impossible it may now seem, an automaton may be contrived which will have its "own mind," and by this I mean that it will be able, independent of any operator, left entirely to itself, to perform, in response to external influences affecting its sensitive organs, a great variety of acts and operations as if it had intelligence.

Three features of Tesla's vision should be highlighted. First, he envisioned that mechanical beings would, over time, bear an increasing resemblance to biological organisms, and in particular human beings. Second, he says that his invention "was evolved", implying that his inventions represent another stage in the process of evolution. And third, the gap between man and automaton will be bridged less by making automata more biological than by recognizing that we biological organisms are automata.

Turning his gaze towards himself, Tesla observes, "I have by every thought and act of mine demonstrated, and do so daily to my absolute satisfaction, that I am an automaton endowed with the power of movement, which merely responds to external stimuli beating upon my sense organs, and thinks and acts and moves accordingly. ..." Tesla would later use the term "meat machines" to refer to human beings, saying that we are simply finer automata than we have yet managed to invent.

ABOVE. 1939 poster for Karel Čapek's *R.U.R.*, which helped to popularize the term "robot."

OPPOSITE. The wreckage of the USS *Maine* in Havana Harbour in 1898. The sinking of the *Maine* helped to propel the US into the Spanish-American War.

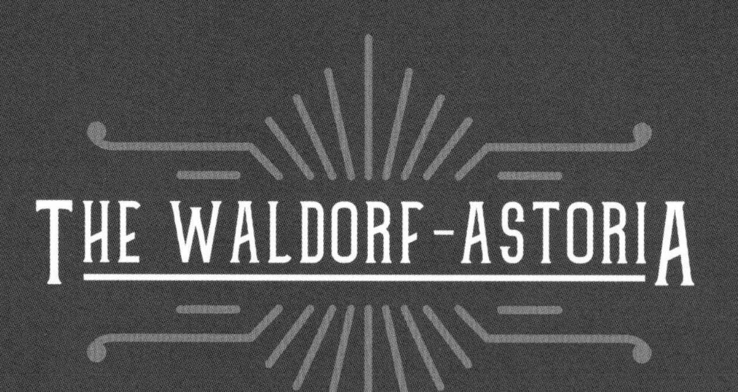

THE WALDORF-ASTORIA

Tesla's best-known residence was the Waldorf-Astoria hotel, where he lived off and on for approximately two decades beginning in 1898. Located in New York City's Midtown Manhattan, the hotel was built in two stages, the Waldorf opening in 1893 and the Astoria in 1897. They were located at the intersections of Fifth Avenue and 33rd and 34th streets. It is impossible to visit these buildings today because they were demolished in 1929 to make way for the Empire State Building.

ABOVE. Engravings of the original Waldorf and Astoria hotels, where Tesla lived for about 20 years.

OPPOSITE. Photograph from a 1909 Waldorf-Astoria banquet for Elbert Gary, founder of US Steel.

The hotels were built on family property by feuding factions of the wealthy Astor family, who had immigrated from Walldorf, Germany. When the Waldorf opened, its prospects for success appeared bleak. People ridiculed its luxury, its impractically high prices for business travellers, and its obtrusive presence in the desirable neighbourhood that housed it. However, its developer staged a benefit concert that attracted the city's elite, and soon the "palace", as the *New York Times* called it, prospered.

The *New York Sun* (15 March 1893) wrote of the new hotel's opening:

> *To American enterprise is due most of the movement toward luxurious hotels which exists in all parts of the world to-day. ... In few palaces of the Old World can such wealth of costly and artistic surroundings be found. The fashionable people ... found private suites of rooms, private dining-rooms, salons, bedrooms such as kings could not excel. ... tapestries and paintings, frescoings, woodcarvings ... marble and onyx and mosaics ... quaint and rich pieces of furniture ... rare and costly tableware. ... the people moving about came upon more wonders than could be seen in one evening.*

When the Astoria opened four years later, it exceeded its predecessor's standards, featuring private bathrooms, a telephone in every room, and most interestingly from Tesla's perspective, full electrification. Though a separate building, the Astoria was constructed so that it could be connected to the Waldorf, which was accomplished with the construction of the colonnade "Peacock Alley". Combined, the Waldorf-Astoria became known as the largest hotel in the world.

The Astoria was seventeen stories high and featured "perpendicular railways" (elevators), a rooftop garden, and a grand ballroom seating 1,500. Its manager, George Boldt, introduced "room service", which enabled guests to dine in bed, and installed an orchestra in the hotel lobby. Added to the Waldorf's 450 rooms, each with its own exterior window, the Waldorf-Astoria boasted 1,000 rooms, as well as three floors of banquet and meeting rooms.

One of the most shining symbols of the United States' "Gilded Age", the Waldorf-Astoria served New York's most notable visitors, including kings and queens, heads of state, and titans of industry. Tesla made sure he cut a corresponding figure. Outfitting his slim,

ABOVE. Oscar Tschirky, the "Oscar of the Waldorf", photographed in around 1885.

OPPOSITE. The Palm Garden dining room at the Waldorf-Astoria Hotel in 1902.

OSCAR TSCHIRKY

One of the hotel's most notable features was its maître d'hôtel, Oscar Tschirky, who worked there for 50 years. He was the former maître d' of Delmonico's, one of New York's best-known restaurants and a Tesla favourite. An immigrant from Switzerland, Tschirky authored a best-selling cookbook in 1896 that included his recipes for Waldorf salad and thousand island dressing. He supervised a staff of 1,000 people and ran a school for waiters, the only one of its kind in the US.

In 1898, the *New York Sun* praised the famous "Oscar of the Waldorf" in these terms:

> *In only one New York hotel ... is there a personage deserving to be called a maître d'Hotel. Anyone who studies him closely will soon arrive at a firm conviction that he might quite as appropriately have been called General or Admiral, if circumstances had not led him into the hotel business. Oscar knows everybody.*

six-foot two-inch frame in coat, tails, cane, top hat and white gloves, the famous inventor dined, usually alone, at 8:10 pm every day. It is said that the germophobic Tesla's table featured two-dozen napkins, which he used to wipe down every dish and utensil.

Despite his eccentricities, Tesla enjoyed a superb reputation as a host, and when he held dinner parties, he often took his guests back to his laboratory, treating them to displays of his electrical wizardry. After the guests departed for the evening, Tesla would typically resume his work until as late as 3:00 am. Though fawned over by a number of society women, Tesla never married, claiming that he felt unworthy and protesting that his dedication to his work would preclude sufficient attention to his spouse.

It is possible that Tesla was attracted to the Waldorf and its social set by a desire to prove his success to himself and others. But it is more likely that Tesla sought out the rich and famous as a means of advancing his investigations. The work he planned to do required substantial financial backing, and such support was not likely to be forthcoming if people of means did not know, respect and perhaps even revere the man undertaking it.

MARCONI

The importance of patents is enshrined in the US Constitution, which reads, "To promote the progress of science and useful arts, by securing for limited time to authors and inventors the exclusive right to their respective writings and discoveries. ..." Without the granting of such patents, the framers of the Constitution feared, a powerful incentive for progress would be missing, slowing the rate at which ideas and inventions enrich human life.

One of the most celebrated patent disputes in US history involved Tesla and the Italian inventor Guglielmo Marconi. Both made important contributions to the development of what came to be known as radio, but the lion's share of the credit and wealth associated with it went to Marconi, who shared the Nobel Prize in Physics for his work in 1909. Though both men were geniuses, most of the entrepreneurship and business acumen was also ascribed to Marconi.

Guglielmo Marconi was born in the university town of Bologna, Italy in 1874. His father was a landowner and a nobleman, and his mother was of Scottish/Irish descent. He and his brother spent a large portion of their childhood in Britain, but in his teens he was allowed to attend lectures at the University of Bologna. There he developed an interest in electricity, and by the 1890s he was hard at work on the problem of "wireless telegraphy".

Many people, including Tesla, had attempted to crack this nut before, but none had managed to produce a commercially viable method of communication. Marconi experimented with radio waves in his attic laboratory, and one night demonstrated to his mother a system that could make a bell ring on the opposite side of the room. Continually refining his apparatus, he eventually managed to transmit signals at distances of up to half a mile.

In 1895 he dramatically expanded the range of his device by grounding the device and using longer antennas. Unable to interest the Italian authorities in his work, he travelled to England, where it met with a much more enthusiastic response. Over the following several years, Maroni demonstrated that he could transmit signals over several miles, over the open sea, and in 1899, across the English Channel. That same year, he conducted the first US demonstrations of his system.

Marconi also proved that signals could be transmitted and received by seafaring vessels, and in 1901 he sent radio signals across the Atlantic Ocean. He discovered that radio signals travel much further at night than during the day, a phenomenon that also

ABOVE. Guglielmo Marconi, the Italian inventor often referred to as the "father of radio".

LEFT. Plaque on the BT Centre, London, commemorating Marconi's first public transmission of wireless signals in 1896.

RMS *TITANIC*

Marconi's fame was dramatically enhanced by the sinking of the RMS *Titanic*, which occurred in 1912. One of those who died was Jack Astor, who built Manhattan's Astoria hotel. While radio did nothing to avert the ocean liner's collision with an iceberg, it did enable a distress signal to reach another vessel, which responded in time to rescue many survivors. Some years later, David Sarnoff, who became the president of the Radio Corporation of America (RCA), claimed that he personally maintained communications with the rescue vessel for 72 hours.

Whether Sarnoff's claim was true or not, radio certainly played a crucial role in the rescue.

Ironically, Marconi had been offered free passage aboard the *Titanic* on its maiden voyage, but had elected to travel on another ocean liner. Later, during a Court of Inquiry on the disaster, Marconi testified about the role that wireless telegraphy had played and could play in the future. A British official later stated, "Those who have been saved have been saved through one man, Marconi."

BELOW. The RMS *Titanic* departing on its maiden voyage in April 1912.

helps to explain why radio signals can travel much further than line of sight. The key is the ionosphere, an upper layer of the atmosphere that reflects short-wavelength radio waves and is less ionized at night, when the sun is not shining on it.

Marconi filed his British patent for radio in 1897, which utilized the discoveries and techniques of a number of predecessors. Later that year, he founded the Marconi Company, which began outfitting both land- and ship-based stations with radio equipment.

LEFT. David Sarnoff, an early wireless telegraph operator of the Marconi Company of America who went on to become the chair of RCA, one of the most important manufacturers of radios and later televisions.

BELOW. Marconi type 31 crystal receiver radio equipment for ships. Crystal radio receivers were first built around 1900, and used crystalline minerals to detect radio signals.

Early in the twentieth century, the US Patent Office also awarded Marconi a patent for radio, and his reputation as the inventor of radio seemed secure with the awarding of the 1909 Nobel Prize.

Tesla had laid much of the groundwork for radio and successfully demonstrated its operation in radio-controlled devices, but he failed to transform his innovations into commercially successful products. He demonstrated in 1893 how a system of radio communication might work, and he obtained radio patents in 1900. When informed of Marconi's work, Tesla responded, "He is a good fellow. Let him continue. He is using seventeen of my patents."

By the time Marconi was awarded the Nobel Prize, he was a rich man, while Tesla's resources were quite limited. Lacking the wherewithal to sustain lengthy and expensive patent litigation, Tesla could do little to overturn Marconi's reputation as the "father of radio". When, years later, the Marconi company reasserted its patent rights in the US, the Supreme Court ruled in 1943 that Marconi was not the first person to conceive of radio, and that the first radio patent belonged to Tesla.

By then, however, both Marconi and Tesla were dead. Marconi had died six years before, following a series of heart attacks. His passing was marked by an Italian state funeral, and all BBC transmitters observed two minutes of radio silence in his honour. Though Marconi relied on the prior work of a number of physicists and inventors, he was the first to transmit and receive radio signals over long distances and founded what became one of Britain's most successful manufacturers of radio equipment.

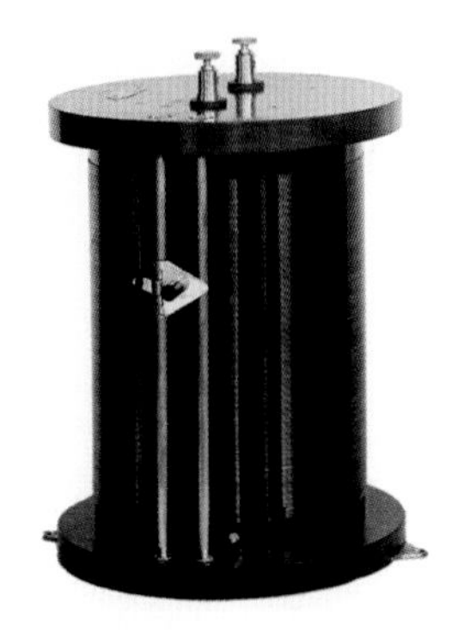

COLORADO SPRINGS

In New York, Tesla managed to convince John Astor to supply funding for his research to the sum of $100,000, though Astor supplied only the first $30,000. It seems that Astor believed he was funding Tesla's investigations into wireless lighting, but Tesla was pursuing a still bigger prize – namely, the planet-wide transmission of power and communications using the earth itself – not wires – as the medium.

To conduct experiments on such a large scale, Tesla could not remain in the crowded city of New York, where laboratory space was expensive to come by, a fire could have catastrophic consequences, and the hustle and bustle of the city made it difficult to tune a sensitive circuit. A lawyer who had represented both Westinghouse and Tesla in patent disputes during the War of the Currents (see p.42) wrote to Tesla that Colorado Springs could supply his needs. There, his associate assured him, both land and electricity would be free.

So, in 1899, Tesla relocated to Colorado Springs, at the foot of Pikes Peak in the Rocky Mountains. Upon his arrival, he revealed to a reporter his intention to transmit a message from Pikes Peak to Paris during the World Exposition in Paris in 1900. He engaged a local builder and began construction on what came to be known as the Tesla Experimental Station. Perhaps as much to incite interest as to protect the curious, he posted warning signs around the site. Borrowing from Dante's *Divine Comedy*, one read, "Abandon Hope, All Who Enter Here".

The station was a large building that resembled a barn. From within rose a tower that extended 37 metres (122 feet) into the air, topped by a copper ball 1 metre (3 feet) in diameter. Equally important, since Tesla planned to use the earth itself as a transmission medium, would be the substructure of the station. Tesla had a 6-metre (20-foot) copper plate buried 3.6 metres (12 feet) below, and he added additional water in an attempt to improve the conductance of the naturally dry soil.

As work progressed over the summer, a huge electrical storm moved through Colorado Springs. Tesla reported that he was able to observe no fewer than 10,000 bursts of lightning in the short space of two hours. He convinced himself that his instruments were able to detect such discharges even long after the storm had passed from sight, which seemed to confirm that such currents could be passed through the earth even at great distances.

Tesla also set up massive oscillators and coils capable of producing electrical discharges at energies ranging up into the millions of volts. With such an apparatus, he could produce artificial lightning bolts up to 41 metres (135 feet) long, whose thunderous reports could be heard as far away as 24 kilometres (15 miles). The local people who lived within walking distance of the station were fascinated with Tesla's work, which seemed to be harnessing the most elemental and powerful forces of nature to the will of man.

One night while Tesla was in the midst of such work, his apparatus suddenly blacked out. He thought initially that his assistant had turned it off, but it turned out that the huge amounts of current he was drawing had burned out the El Paso county generator, plunging the city and its environs into darkness. The manager of the plant was not pleased, and Tesla was able to mend fences only by taking a crew to the plant and making the necessary repairs himself.

Tesla regularly observed the appearance of ball lightning, which he referred to as "fireballs". Still incompletely understood, ball lightning appears as a spherical, luminous apparition floating in air that can range in size from a few millimetres to several metres in diameter and persist for as long as several seconds. Though Tesla claimed to be able to produce ball lightning at will, from his point of view the

ABOVE. Tesla's Colorado Springs Station, fronted by a sign reading, "Great Danger, Keep Out". The inventor can be seen peeking around the edge of the doorway.

OPPOSITE. Tesla's Colorado Springs Experimental Station. Note the copper ball at the top of the antenna.

phenomenon represented a mere curiosity, and he focused his attention elsewhere.

In one particularly harrowing event at the station, Tesla nearly lost his life. Having sent his assistant into town, he was working alone when, as he recalled,

> *I went behind the coil to examine something. While I was there, the switch snapped in, and suddenly the whole room was filled with [electrical] streamers, and I had no way of getting out. I tried to break through the window but in vain, as I had not tools and there was nothing else to do than to throw myself on my stomach and attempt to pass under. The primary carried 50,000 volts, and I had to crawl through a narrow place with the streamers going. The nitrous acid was so strong I could hardly breathe. ... When I came to the narrow space they closed on my back. I got away and barely managed to open the switch when the building began to burn. I grabbed a fire extinguisher and succeeded in smothering the fire.*

Perhaps the most famous photograph of Tesla was taken during his work in Colorado Springs. He had arranged for a photographer to document his most interesting work at the station, and one day the photographer captured an image of lightning-like discharges streaming across the room from Tesla's magnifying transmitter. On this he superimposed a photo of Tesla sitting quietly and reading. The multiple exposures made it appear that Tesla was sitting calmly in the midst of a lightning storm.

OPPOSITE ABOVE. A series of photographs of ball lightning.

OPPOSITE BELOW. A portion of the oscillator used in Tesla's Colorado Springs laboratory.

BELOW. Multiple-exposure photograph of Tesla and his magnifying transmitter.

COPYING AND ENLARGING

Lantern Slides
Colored and Uncolored

INTERIORS and ANIMALS
Artistically Fotografed.

SPECIAL FACILITIES FOR Amateur Developing and Finishing.

Colorado Springs, Colo., Nov 3rd 1899

Nicola Tesla

IN ACCOUNT WITH....

THE STEVENS FOTOGRAFERIE

24-26 EAST BIJOU STREET.

ALL ON GROUND FLOOR.

FINEST COLORADO VIEWS,
COLORED AND UNCOLORED.

PORTRAIT SITTINGS
In charge of High-Class ARTISTS.

FOTO NOVELTIES
In Great Variety.

Sept	21	To 2 hrs time @ 60¢ per hr.	1	20
"		To 5 6x8 Negatives @ 75¢ ea.	3	75
"	23	To 3 hrs time @ 60¢	1	80
"	"	To 10-6x8 Negatives @ 75¢ ea.	7	50
"	30	To 11 Prints-6x8 @ 75¢ ea.	8	25
Oct	9	To 7 Hours time @ 60¢	4	20
"	"	To 14-6x8 Negatives @ 75¢ ea	10	50
"	"	To 1/4 Lb. Flash Light Powder	2	00
"	10	To 11 Hours time @ 60¢	6	60
"	"	To 5g Flash Light Powder	2	50
"	"	To 8 - 8x10 Negatives @ $1.00 ea	8	[illegible]
"	"	To 9 6x8 " @ 75¢ ea	6	75
"	10	To 2 Platinum 6x8 Pts - @ 75¢ ea	1	50
"	13	To 20 - 14x17 Prints Double Mtd @ $5.00 ea	100	00

ABOVE. One of the receipts for photographs taken at the Colorado Springs laboratory, which included some of the most famous images ever obtained of Tesla and his work.

Colorado Springs, Colorado, Dec 1 1899

Nikola Tesla

To The El Paso Electric Co., Dr.

Folio

Oct	31	Time and labor for linemen and helpers putting up new circuit	79 99
		57200 lbs Coal used during Oct 143 hrs	47 90
		3 months light 1 arc lamp	27 00
Nov	1	Carfare for linemen	35
	16	Changing converter	20
		19-16 CP lamps	4 75
	30	72800 lbs Coal used during Nov 182 hrs	54 60
		Oil & Waste	5 00
		2 arc lights for Nov	18 00
		Time & labor changing converters and burned out armature	$ 7 57
			245 36

EL PASO ELEC. CO.
PAID
DEC 7 1899
By Davies

ABOVE. Receipt for parts and labour needed to repair damage caused by Tesla's experiments at the El Paso power station in Colorado Springs, 1899.

TALKING WITH THE PLANETS

Around the time Tesla was conducting experiments in Colorado Springs, Marconi was providing public demonstrations of his ability to transmit signals wirelessly over ever greater distances. As Tesla expressed it, Marconi was merely modifying existing discoveries to improve their performance, a menial kind of work that Tesla professed to find uninteresting. By contrast, Tesla was pursuing new discoveries that could revolutionize wireless communication, virtually eliminating the problem of distance.

It was late in the year and while alone at night in the experimental station that Tesla detected a phenomenon so remarkable that he quickly convinced himself it would eclipse his work on wireless communication. His own words from a February 1901 article in *Collier's Weekly*, "Talking with the Planets", capture what happened:

> *As I was improving my machines for the production of intense electrical actions, I was also perfecting the means for observing feeble effects. One of the most interesting results, and also one of great practical importance, was the development of certain contrivances for indicating at a distance of many hundred miles an approaching storm, its direction, speed, and distance traveled. ... It was in carrying on this line of work that for the first time I discovered those mysterious effects which have elicited such unusual interest. ... I can never forget the first sensations I experienced when it dawned upon me that I had observed something possibly of incalculable consequences to mankind. ... My first observations positively terrified me, as there was present in them something mysterious, not to say supernatural, and I was alone in my laboratory at night. ... The changes I noted were taking place periodically, and with such a clear suggestion of number and order that they were not traceable to any cause known to me. ... It was some time afterward when the thought flashed upon my mind that the disturbances I had observed might be due to an intelligent control. ... The feeling is constantly growing on me that I had been the first to hear the greeting of one planet to another.*

What Tesla had detected alone in his laboratory that night was a series of electrical signals that seemed to represent a regular numerical progression, "one, two, three". He considered the possibility that they might represent some natural phenomenon, such as solar activity or the aurora borealis (northern lights), but he quickly excluded such possibilities. To Tesla, these

OPPOSITE. 1901 illustration of the Wardenclyffe Tower on Long Island, as Tesla hoped it would eventually appear.

WORLD WIRELESS
RETURN
TO
NIKOLA TESLA CO.
8 West 40 St., N.Y.

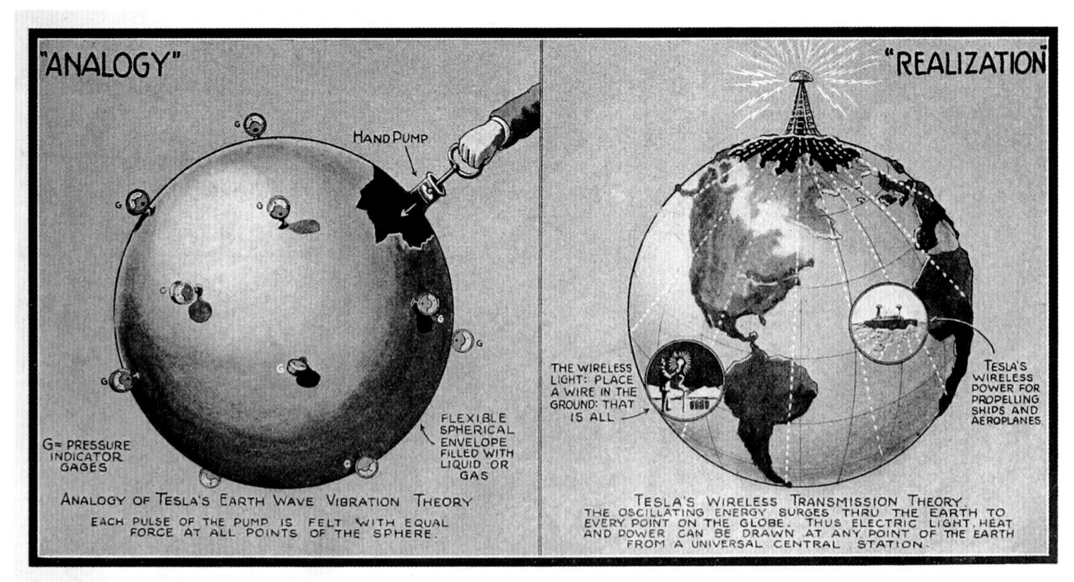

Tesla's World-Wide Wireless Transmission of Electrical Signals, As Well As Light and Power, Is Here Illustrated in Theory, Analogy and Realization. Tesla's Experiments With 100 Foot Discharges At Potentials of Millions of Volts Have Demonstrated That the Hertz Waves Are Infinitesimal In Effect and Unrecoverable; the Recoverable Ground Waves of Tesla Fly "Thru the Earth". Radio Engineers Are Gradually Beginning to See the Light and That the Laws of Propagation Laid Down by Tesla Over a Quarter of a Century Ago Form the Real and True Basis of All Wireless Transmission To-Day.

signals bore the unmistakable marks of "intelligent beings" – probably waiting for a response such as "four, five, six".

With the nineteenth century drawing to a close, Tesla was asked by the American Red Cross for a prediction regarding the greatest achievements the world would witness in the next century. In a letter dated "Christmas", perhaps because Tesla saw himself as presenting mankind with its most remarkable gift ever, he wrote:

The retrospect is glorious, the prospect is inspiring: Much might be said of both. But one idea dominates my mind. This – my best, my dearest – is for your noble cause. I have observed electrical actions, which have appeared inexplicable. Faint and uncertain though they were, they have given me a deep conviction and foreknowledge, that ere long all human beings on this globe, as one, will turn their eyes to the firmament above, with feelings of love and reverence, thrilled by the glad news: "Brethren! We have a message from another world, unknown and remote. It reads: one ... two ... three ..."

Though Tesla did not explicitly attribute these signals to Martians or the inhabitants of any other particular planet, he did much to fan the flames of public speculation. For example, in his 1901 article in the *Colorado Springs Gazette*, he wrote:

At the present stage of progress, there would be no insurmountable obstacle in constructing a machine capable of conveying a message to Mars, nor would there be any great difficulty in recording signals transmitted to us by the inhabitants of that planet, if they be skilled electricians. ... What a tremendous stir this would make in the world! How soon will it come? For that it will some time be accomplished must be clear to every thoughtful being.

However, Tesla's reports did not meet with universal approbation. On the contrary, a number of academics were outraged that a man who had made so many useful scientific and technological contributions should engage in such fanciful and unsupported speculations. One professor at the University of California described Tesla's folly in these terms:

It is the rule of a sound philosophizing to examine all probable causes for an unexplained

The Colorado Springs Gazette January 9, 1901

NIKOLA TESLA AND HIS TALK WITH OTHER WORLDS

Special to the Philadelphia Inquirer.

New York, Jan. 2.—Not quite two years ago, Mr. Nikola Tesla went out to Colorado to conduct experiments in relation to the wireless transmission of energy which has engaged his attention for several years.

Mr. Tesla dwelt on his work to an Inquirer man this afternoon. He regards his latest results as the most important he has ever attained. Briefly Tesla has been able to note a novel manifestation of energy which he knows is not of solar or terrestrial origin, and being neither, he concludes that it must emanate from one of the planets.

While he was conducting his investigations in his Colorado laboratory the instrument he was using to observe the electrical condition of the earth was affected in an unaccountable manner. It recorded three distinct, though very faint, movements, one after the other.

May Talk With Other Worlds.

These movements were observed not once but many times, the number of impulses varying, and Mr. Tesla now firmly believes that with improved apparatus it will be quite possible for the people of the earth to communicate with the inhabitants of other planets. In tetlling about his work and his discoveries, Mr. Tesla said:

"I set out to carry on my experiments along three different lines: first, to ascertain the best conditions for transmitting power without wires; second, to develop apparatus for the transmission of messages across the Atlantic and Pacific oceans, on which problem I have been engaged for eight years, and third, to work on another problem which involves a still greater mastery of electrical forces, and which, with my present knowledge, I consider of still greater importance than even the transmission of power without wires, and which I shall make known in due course.

Pressure of 50,000,000 Volts.

"There were, however, numerous points to be found out about electrical vibrations, and there were actions on which I was still in doubt. In my laboratory in New York I was able to go only to electrical discharges of 16 feet in length, and I had only reached effective electrical pressures of about 8,000,000 volts. To carry the problems on which I was working farther I had to master electrical pressures of at least 50,000,000 volts, and electrical discharges were necessary for some purposes measuring at least 50 or 100 feet.

"The results I attained were far beyond any I had expected to reach. One of the first observations I made in Colorado was of great scientific importance, and confirmatory of a result I had already obtained in New York. I refer to my discovery of the stationary electrical waves in the earth. The significance of this phenomenon has not yet been grasped by technical men, but it virtually amounts to a positive proof that with proper apparatus such as I have perfected a wireless transmission of signals to any point on the globe is practicable.

Sparks One Hundred Feet Long.

"In perfecting my apparatus I encountered at first great difficulties. I had a few narrow escapes from sudden sparks jumping out to great distance and a number of times my laboratory caught fire, but I carried all the work through without a serious mishap. I gradually learned how to confine electrical currents of a pressure of 50,000,000 volts to produce electrical movements up to 110,000-horse power, and I succeeded in obtaining electrical discharges measuring from end to end 100 feet and more. These results were, however, rendered more valuable by the fact that they opened up still greater possibilities for the future.

"It was in investigating feeble electrical actions transmitted through the earth that I made some observations which are to me the most gratifying. Chief among these certain feeble electrical disturbances which I could barely note occurred, and which by their character unmistakably showed that they were neither of solar origin nor produced by any causes known to me on the globe. What could they be?

"I have incessantly thought of this for months, until I finally arrived at the conviction, amounting to almost knowledge, that they must be of planetary origin. As I think over it now it seems to me that only men absolutely stricken with blindness, insensible to the greatness of nature, can hold that this planet is the only one inhabited by intelligent beings.

Communication With Mars.

"I have perfected my transmitting apparatus so far that I can undertake to construct a machine which will without the least doubt be fully competent to convey sufficient energy to the planet Mars to operate one of these delicate appliances which we are now using here, as, for instance, a very sensitive telephone instrument.

"With regard to my work in other lines which I have simultaneously carried on my progress has been most satisfactory, and I hope that soon electrical energy may be turned to the usages of man in a way and for purposes such as to surpass in importance all that we have ever done heretofore."

phenomenon before invoking improbable ones. Every experimenter will say that it is almost certain that Mr. Tesla has made an error, and the disturbances in question come from currents in our air or in the earth. ... Until Mr. Tesla has shown his apparatus to other experimenters and convinced them as well as himself, it may safely be taken for granted that his signals do not come from Mars.

Several hypotheses have been advanced to explain the signals that Tesla detected. One is that the signals were generated by Marconi during his European experiments, though Marconi seems to have been operating in a different frequency range. Or Tesla may have detected signals from another experimenter. Another possibility is that the signals truly emanated from an extraterrestrial source, but

OPPOSITE. Illustration of how Tesla imagined power could be transmitted wirelessly through the earth, in a manner analogous to pumping air into a sphere.

ABOVE. Article from the 9 January, 1901 *Colorado Springs Gazette* describing Tesla's vision of communicating with other worlds.

represented natural phenomena such as interactions between Jupiter and one of its moons.

One thing is certain. Tesla's speculation that the signals he detected originated from an extraterrestrial intelligence did much to damage his standing in the scientific community. Tesla always seemed to relish the role of showman, but at least when he was demonstrating his Tesla coils or his radio-controlled boat, he was operating on sound scientific and technological footings. Now, however, he seemed to be speculating so recklessly that he undermined his own credibility.

In a January 1905 article in *Electrical World and Engineer*, Tesla responded to detractors who ridiculed his more radical plans and speculations in these terms:

Perhaps it is better in this present world of ours that a revolutionary idea or invention instead of being helped and patted, be hampered and ill-treated in its adolescence – by want of means, by selfish interest, pedantry, stupidity and ignorance; that it be attacked and stifled; that it pass through bitter trials and tribulations, through the heartless strife of commercial existence. ... So all that was great in the past was ridiculed, condemned, combated, suppressed – only to emerge all the more powerfully, all the more triumphantly from the struggle.

Tesla returned to New York with even grander plans.

OPPOSITE. Mars, the "Red Planet", second-smallest in the solar system, and mentioned by Tesla as the possible source for what he took to be extraterrestrial radio signals indicative of intelligence.

BELOW. One of Tesla's early experiments using an early conical alternator, attempting wireless energy transmission, 1894.

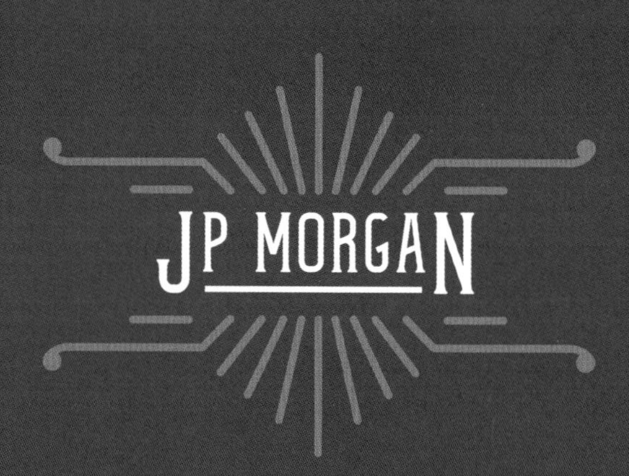

JP MORGAN

One of the wealthiest and most influential men of his era, the founder of many of the industrial giants of the late nineteenth and early twentieth centuries, John Pierpont Morgan was born into wealth in Hartford, Connecticut in 1837. His father, Junius Morgan, was a highly successful banker who groomed his son to take over the family business, ensuring that he spent time in both Switzerland and Germany to improve his proficiency with foreign languages.

A sickly child who suffered from seizures, Morgan grew into a large and powerfully built man. He started his banking career with his father's friends in London, but soon moved to New York. Avoiding service in the Civil War by paying a substitute to take his place, Morgan profited from selling arms at a considerable mark-up. He became a partner in dozens of firms over the years, and developed a reputation for taking charge, reorganizing many to profitability through a process that became known as "Morganization".

Morgan and Thomas Edison crossed paths in the 1880s. Edison had interests in numerous electric companies, including manufacturers of lamps, motors and electrical fixtures. In 1889, Morgan's company provided financing for Edison's research and merged a number of these firms into the Edison General Electric Company. This company then merged with the Thomson-Houston Electric Company to form General Electric, which became one of the original 12 companies in the Dow-Jones Industrial Average, the US stock market index founded in 1895.

Morgan was a notoriously taciturn man, very effective in action but not inclined to offer much in the way of explanation. For example, when he was asked what the stock market would do, he replied simply, "It will fluctuate." In another case, he was asked why he had become involved in the purchase of a life insurance company whose securities yielded only one-eighth of 1 per cent. Replied Morgan, "For the very reason that I thought it was the thing to do".

Morgan also professed to uphold his father's deep conviction that character is everything. In Congressional testimony the year before his death, Morgan declared:

> *The first thing is character, before money or anything else. Money cannot buy it. A man I do not trust could not get money from me on all the bonds in Christendom. I think that is the fundamental basis of business.*

OPPOSITE. John Pierpont Morgan (1837–1913), one of the titans of American industry and finance, as photographed in 1910.

FIELD MARSHAL OF INDUSTRY

When a panic in the mid-1890s nearly exhausted the US Treasury's gold supply, Morgan stepped into the breach, arranging for the government to purchase gold from Morgan and his banking associates. At the turn of the century, Morgan set his sights on consolidating the steel industry, the biggest prize of which was Andrew Carnegie's eponymous company. Morgan asked how much it would cost to buy the business. Carnegie wrote down "$480 million". Morgan glanced at the paper and said, "I accept this price." The result was US Steel, the first billion-dollar company.

A second financial crisis developed in 1907, again brought on largely by the US's chronic gold shortage. Facing a run on the banks, Morgan invited the presidents of the leading banks to his home, where he locked the door, letting them know that no one would leave until an agreement was reached. At 4.00 am the next morning, Morgan set forth an agreement stipulating the amount that each would pledge to end the crisis. He then pointed to the signature line for each president, telling them, "Here is the pen."

Morgan did much to strengthen the US economy, although most of his efforts also profited him and his associates. He became known as the "field marshal of industry" and the "financial Moses of the New World". The trust-busting US President Theodore Roosevelt railed against Morgan as one of the "malefactors of great wealth", breaking up a railroad trust that Morgan had formed. The US Supreme Court ultimately upheld the decision, stating that the trust restrained trade in violation of the Sherman Anti-Trust Act.

RIGHT. Crowds outside the Federal Hall on Wall Street during the autumn 1907 financial crisis, which undermined Tesla's efforts to raise funds.

RIGHT. A 1911 political cartoon lampooning Morgan's outsize influence on the US economy.

BELOW. Morgan, notoriously self-conscious about his large and deformed nose (the result of a medical condition), was extremely reluctant to be photographed, and is shown here striking a photographer with his cane.

Morgan's impact on American business was immense. In addition to helping to found US Steel and General Electric, he also played important roles in the creation of AT&T and International Harvester. When he died in Rome in 1913 at the age of 75, flags on Wall Street flew at half mast, and when his funeral procession moved through the city of New York, the stock market closed. Soon thereafter the US abandoned the gold standard, and the Federal Reserve was created to take over management of the money supply.

As one of the foremost financiers of his era, it was only natural that in 1900 Morgan should find himself approached by Tesla, who regaled him with the vision of a transatlantic system of wireless communication. Compared to Marconi's system, which at the time operated over much shorter distances, Tesla's system, to be installed at a site he named Wardenclyffe on Long Island, New York, appeared to represent one of the most transformative (and potentially profitable) innovations in the history of communication. Perhaps to his ultimate regret, Morgan agreed to provide Tesla with $150,000.

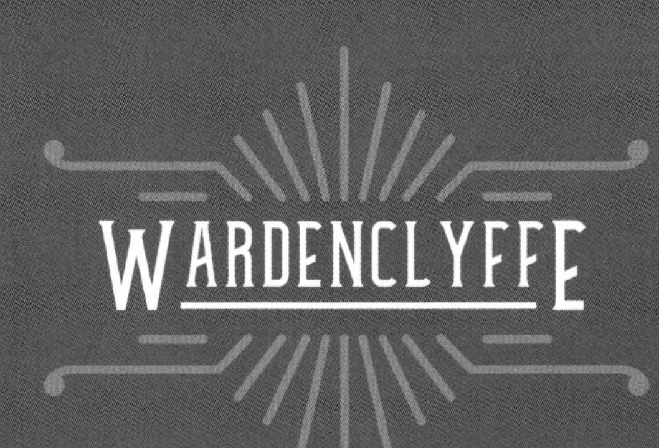

WARDENCLYFFE

Tesla chose the site for his "World Telegraphy Center" based on several criteria. First, he needed a large amount of space, envisioning not merely a transmitter tower but laboratories and manufacturing facilities – something akin to today's industrial parks. In addition, he wanted something secluded, so his work could proceed in secret. Finally, he needed inexpensive land within a train ride of his New York base of operations.

A real-estate developer on Long Island named Warden who held a nearly 810-hectare (2,000-acre) parcel on Long Island offered Tesla 81 hectares (200 acres). Imagining that Tesla's work would generate many jobs, first for those building the facility and later for those who would operate it, the developer expected to supply housing for them.

As soon as the site was selected, Tesla set about purchasing equipment and supplies. He also arranged for his representatives to begin locating a site for a second station on the other side of the Atlantic.

As Wardenclyffe proceeded, Tesla soon needed more funds. He approached Morgan, to whom he had already assigned a 51 per cent interest in the venture and any patents to which his wireless telegraphy might give rise. Tesla first argued that more money

RIGHT. Artist's depiction of the completed Wardenclyffe Tower from *Electrical Experimenter* magazine, 1919.

OPPOSITE. Diagram from US Patent 1,119,732, Tesla's "apparatus for transmitting electrical energy" – the fundamental principle behind Wardenclyffe.

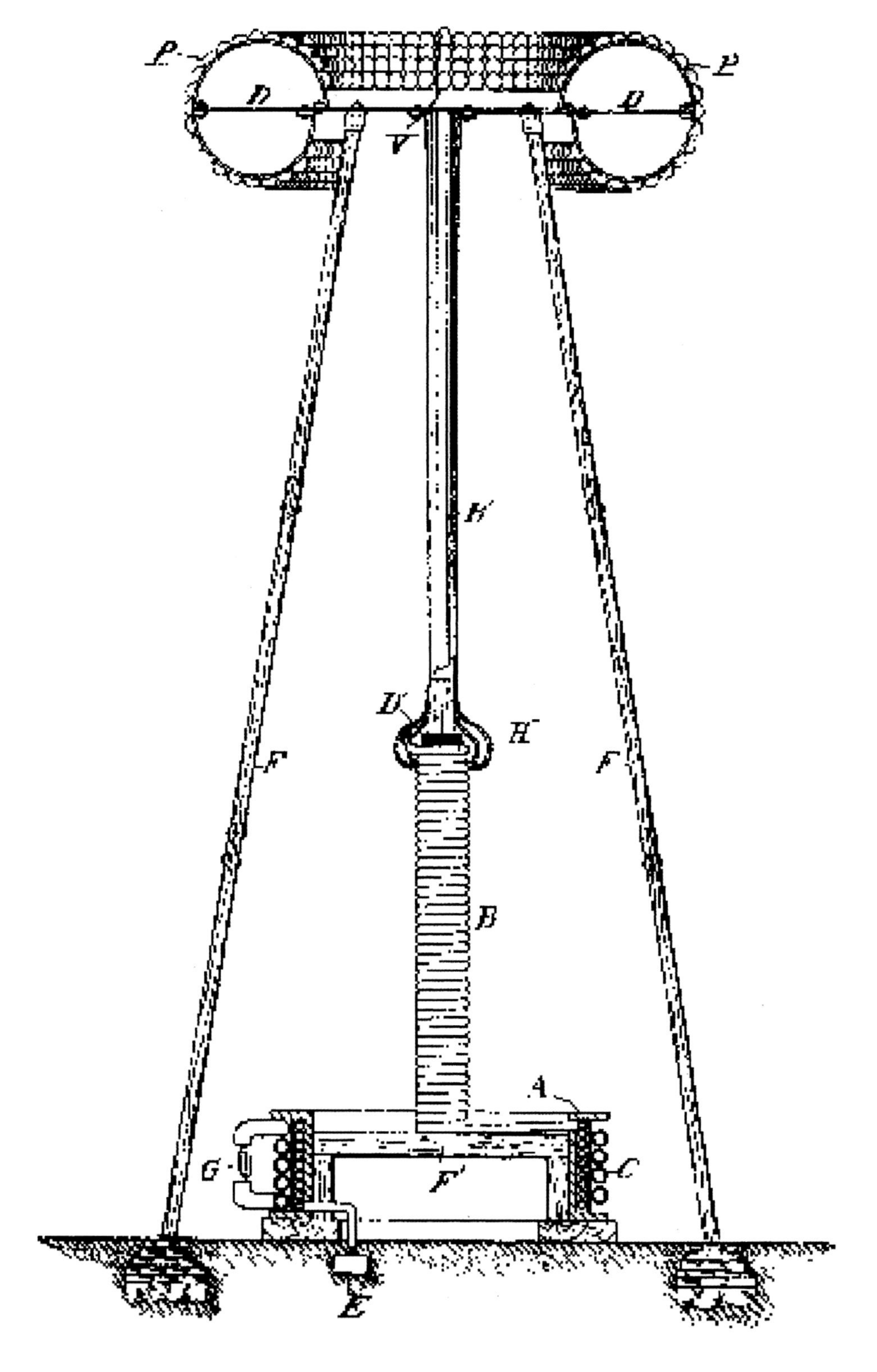
P
P
V
F
F
B
A
G
C
F'
E

STANFORD WHITE

Tesla originally envisioned a tower that would rise 183 metres (600 feet) into the air, but eventually settled for a tower only a fraction of this height. For one thing, the stock market collapsed, and the prospects for additional funding dimmed considerably. In addition, Tesla's architect, Stanford White, convinced him that such a grand design would prove simply too expensive. Feeling the pressure from Marconi's ever-lengthening radio transmissions, Tesla agreed to a scaled-down version.

Just five years later, White himself would come to an unfortunate end. A well-known architect and designer for the rich and famous, White was also a serial philanderer. One of his conquests was a young actress, Evelyn Nesbit, who later married a mentally unstable millionaire, Harry Thaw. One night at a theatrical production, Thaw approached White with a pistol drawn and fired three shots at White, killing him instantly. What followed was widely hailed by the press as the "Trial of the Century". Thaw pleaded temporary insanity and was eventually acquitted and set free.

RIGHT. Front page story on the Stanford White killing, *New York World*, 27 June 1906.

NOTE FROM WHITE TO HIS BRIDE DROVE HARRY THAW TO MURDEROUS REVENGE.

Evelyn Nesbit Thaw Begins a Loyal Fight for Her Husband's Life and, It Is Said, Has Furnished Letters From the Architect That Led to the Roof-Garden Tragedy.

$40,000,000 WILL BACK THE EFFORT TO ESCAPE DEATH CHAIR.

"I Am Not a Fugitive," Says Young Wife—Visits Lawyers and Leaves Over Roofs—Murderer Goes Through Police Routine and Lands in Tombs Cell—Pittsburg Folks and Friends Say He Has Long Been Known as "Crazy Harry."

would allow him to build a tower large enough to transmit signals across not only the Atlantic Ocean but also across the Pacific. Later, he revealed that he envisioned transmitting not only signals but also power across the ocean.

Over a period of five years, Tesla eventually wrote Morgan more than 50 letters, each pleading for more funds, but the financier never advanced him any more money. As a result, Tesla settled on a tower height of approximately 46 metres (150 feet), one quarter of what he originally envisioned.

The eventual height of the tower was 57 metres (187 feet) above ground level, but Tesla also included a subterranean component that extended 36.6 metres (120 feet) below the ground, giving the total apparatus a height of over 93.5 metres (307 feet).

Especially galling to Tesla was the fact that Marconi was attracting many investors, causing his company's stock price to rise handsomely. At the end of 1901, Marconi would announce to the world that he had successfully conducted the first transatlantic radio transmission, from Cornwall to Newfoundland. At this point, however, Tesla remained convinced that he was on the verge of a truly revolutionary breakthrough, while Marconi was merely extending Tesla's earlier work.

Yet Tesla could not raise sufficient funds to move the project forward. In 1903 he did attend the raising of the 50-tonne (55-ton) dome to the tower's top, but he could not afford to plate it with copper, which would be necessary for it to function as planned. Some began to suspect that Morgan had provided funding merely to prevent the development of Tesla's radio patents, though such suspicions were never substantiated. However, news of Morgan's withdrawal quelled the interest of other investors.

A related problem was the fact that Tesla had assigned to Morgan a majority interest in patents that he would develop. This meant in effect that any other investor in the project would be lining Morgan's already bulging pockets, a prospect that would not have appealed to many. What Tesla took to be a gesture of respect and trust – magnanimously assigning the majority interest to Morgan – ended up contributing to Wardenclyffe's downfall.

Tesla had been purchasing much of his equipment from the Westinghouse company, and in 1906 he wrote to its leader to request support:

> *The transmission of power without wires will very soon create an industrial revolution such as the world has never seen before. Who is to be more helpful in this great development, and who will derive from it greater benefits than yourself?*

Eventually, Westinghouse and his heirs would obtain judgements against Tesla for unpaid bills, and Tesla would twice mortgage Wardenclyffe to the Waldorf-Astoria to secure payment of his hotel debts. The tower was

RIGHT. 1904 Photograph of Tesla's Wardenclyffe Tower.

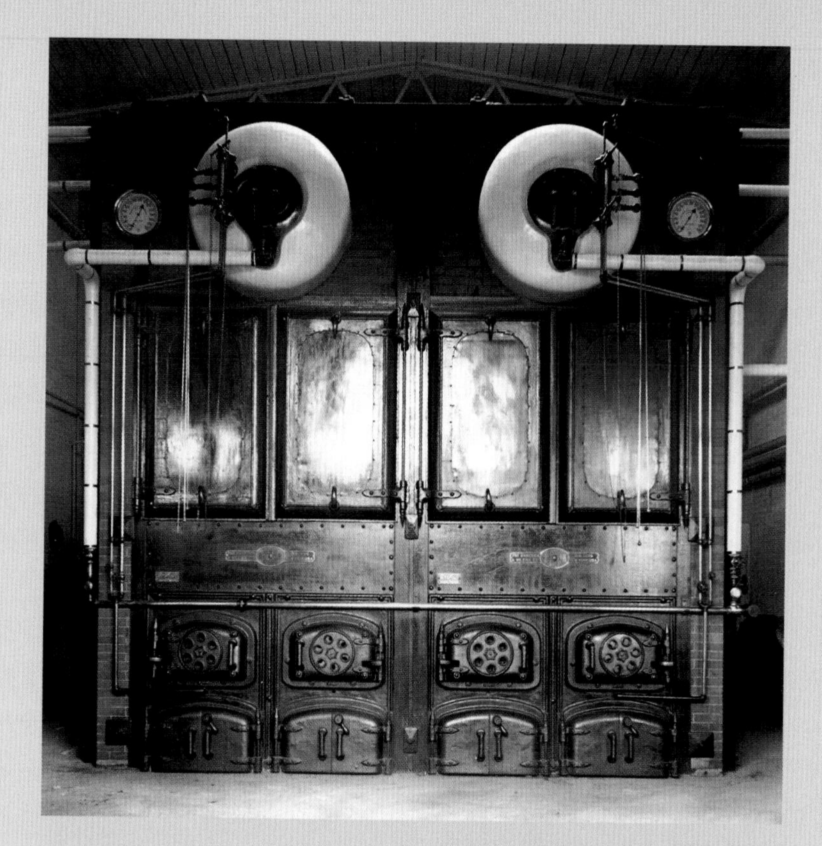

demolished in 1917 and sold for scrap. In a 1968 article entitled "Wardenclyffe: A Forfeited Dream", a *Long Island Forum* reporter described the last days of the facility in these terms:

> *When it became generally known that the Wardenclyffe operation had closed, an occasional research engineer bitten with curiosity would make his way out to the laboratory. If the caretaker was still there, the visitor would be admitted for a tour of the laboratory. What he would behold was something quite beyond his expectations —intricate mechanical mechanisms, glass blowing equipment, a complete machine shop including eight lathes, X-Ray devices, many varieties of high frequency (Tesla) coils, a radio-controlled boat, exhibit cases with at least a thousand bulbs and tubes, an instrument room, electrical generators and transformers, wire, cable, library and office. A strange stillness filled the building. It seemed as if it were a holiday, and the workday tomorrow would bring back all the workers to their assignments. A walk outside to the tower and up the flights of stairs - soon one caught the whisper of the wind through the spars. One believed he could perceive muted voices and clanking sounds below. But the switch had not been thrown. The dynamos stood idle.*

OPPOSITE ABOVE. Equipment at Tesla's Wardenclyffe Tower.

OPPOSITE BELOW. Two-phase generator installed at Wardenclyffe Tower.

RIGHT. The 1917 demolition of Tesla's Wardenclyffe Tower.

Letters to Morgan

To understand what was foremost in Tesla's mind as he struggled to complete the Wardenclyffe project, there is no better source than his own letters to Morgan, which over time grow both more grandiose and more plaintive. In January of 1904 he wrote:

> *Will you help me on any terms you choose and enable me to insure and develop a great property which will ultimately yield hundredfold returns? Please do not do me an injustice in believing me incapable simply because a certain sum of money was not sufficient to carry out my undertaking.*

Tesla, who cared little about money except as a means of achieving his technological dreams, found it almost incomprehensible that something so trivial as a lack of money should stand in his way.

At times, Tesla seems to upbraid Morgan, assigning to him the blame for the fact that the great tower sits incomplete and idle and reproving him for souring prospects of funding from other sources.

Just a day later, he writes:

> *When, after putting in all I could scrape together, I come to show you that I have done the best that could be done, you fire me out like an office boy and roar so that you are heard six blocks away: Not a cent; it is spread all over town. I am discredited, the laughing stock of my enemies.*

From Tesla's point of view, the stakes in this game are not merely financial. They are also reputational, and he suffers mightily each time he imagines his competitors rejoicing at his failure.

By October of 1904, Tesla seems to despair of garnering Morgan's support by describing the great profits his project would produce, and instead begins to describe his own suffering, as if to evoke Morgan's aid through sympathy:

> *Since a year, Mr. Morgan, there has been hardly a night when my pillow is not bathed in tears, but you must not think me a weak man for that. I am perfectly certain to complete my task, come what may.*

Were Morgan to have believed this last assertion, he might have judged his own support not so crucial to the completion of the project as Tesla's pleas seemed to suggest.

That same month, the increasingly frustrated Tesla launched a frontal attack on Morgan, pulling no punches in accusing him of an irrational obstinacy:

OPPOSITE. Promotional illustration for Tesla's plan for worldwide wireless communication.

ABOVE. Wardenclyffe Tower and its framed dome as it appeared in 1904.

You are like Bismarck. Great but uncontrollable. I wrote purposefully last week hoping that your recent association [with the Anglican archbishop] might have rendered you more susceptible to a softer influence. But you are not Christian at all, you are a fanatic [Muslim]. Once you say no, come what may, it is no. ... Will you not listen to anything at all? Are you to let me perhaps succumb, lose an immortal crown? Will you let a property of immense value be depreciated, let it be said that your own judgment was defective, simply because you once said no?

On the one hand, Tesla's sense of urgency to complete the project leads him to address Morgan as though he were its only hope. On the other hand, his personal pride and his disbelief that the will of any mere financier could thwart the achievement of his immortal dream leads him to flaunt other options. For example, he tells Morgan that, should he choose, he could quickly assemble a lecture tour that would net him not less than a few million dollars in just a week.

By year's end, Tesla seems to have become nearly unhinged, implying that Morgan is barely worth his remonstrations and glorifying his own work in the grandest of terms:

You are a big man, but your work is wrought in passing form, mine is immortal. I came to you with the greatest invention of all times. I have more creations named after me than any man that has gone before me not excepting Archimedes and Galileo – the giants of invention). Six thousand million dollars are invested in enterprises based on my discoveries in the United States today.

What to Morgan may have seemed bizarre delusions of grandeur achieved their most feverish pitch in a letter Tesla penned in February of 1905, yet Morgan remained unmoved, and Wardenclyffe sat idle:

Let me tell you once more. I have perfected the greatest invention of all time – the transmission of electrical energy without wires to any distance, a work which has consumed 10 years of my life. It is the long-sought stone of the philosophers. I need but to complete the plant I have constructed and in one bound, humanity will advance centuries. I am the only man on this earth today who has the knowledge and ability to achieve this wonder and another one many not come in a hundred years.

ABOVE. Justus Sustermans' 1636 portrait of Galileo (1564–1642), one of the "giants of invention" in world history in whose company Tesla ranks himself.

OPPOSITE. Interior of the Wardenclyffe Tower.

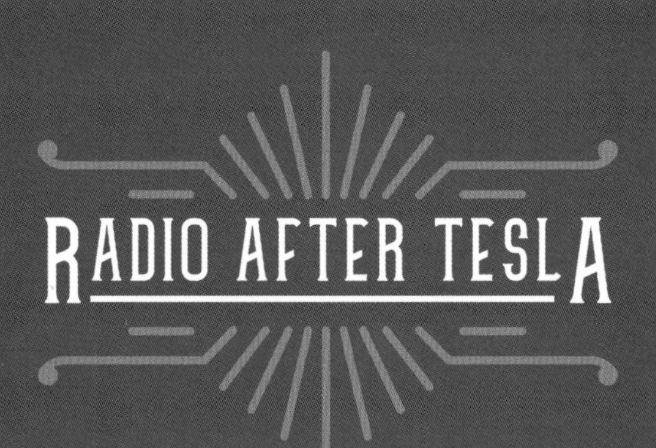

RADIO AFTER TESLA

Tesla had doubts about what we have come to call radio. Known in his day primarily as "Hertzian waves", after the German physicist Heinrich Hertz, who first proved the existence of electromagnetic waves, the very term "radio" reveals the problem. "Radio" comes from the Latin *radius*, meaning beam, and Tesla initially believed that, like light, such beams could never travel further than a line of sight, which would markedly limit their range; hence Tesla's interest in using the earth itself as a medium for signalling and power transmission.

ABOVE. Heinrich Hertz (1857–94), the German physicist who proved the existence of electromagnetic waves.

REGINALD FESSENDEN

Marconi was not the only person who played an important role in developing radio. Another was Reginald Fessenden (1866–1932), who went to work for Edison in the 1880s but was laid off due to the inventor's financial difficulties. After helping to install the lighting for the 1893 Chicago Columbian Exposition (see p.54), he was recruited by Westinghouse as the chair of electrical engineering in Pittsburgh. A company was formed to support his electric signalling research, and he developed a rotary-spark transmitter that enabled him to conduct the first two-way transatlantic radio exchange.

In 1904 Fessenden helped to engineer a Niagara Falls power plant for Ontario, but in 1911 he ceased his radio research after he was dismissed from his company. He then developed what became a type of sonar system for use in enabling submarines to avoid objects such as icebergs and to signal each other. Fessenden eventually accumulated more than 500 patents and in the 1930s claimed that, in 1906, he had been the first person to broadcast entertainment by radio. He also sued RCA for failing to compensate him for some of his patents, eventually winning a large settlement.

ABOVE. Reginald Fessenden, one of the early radio pioneers, in 1906. He may have been the first person to broadcast music.

LEE DE FOREST

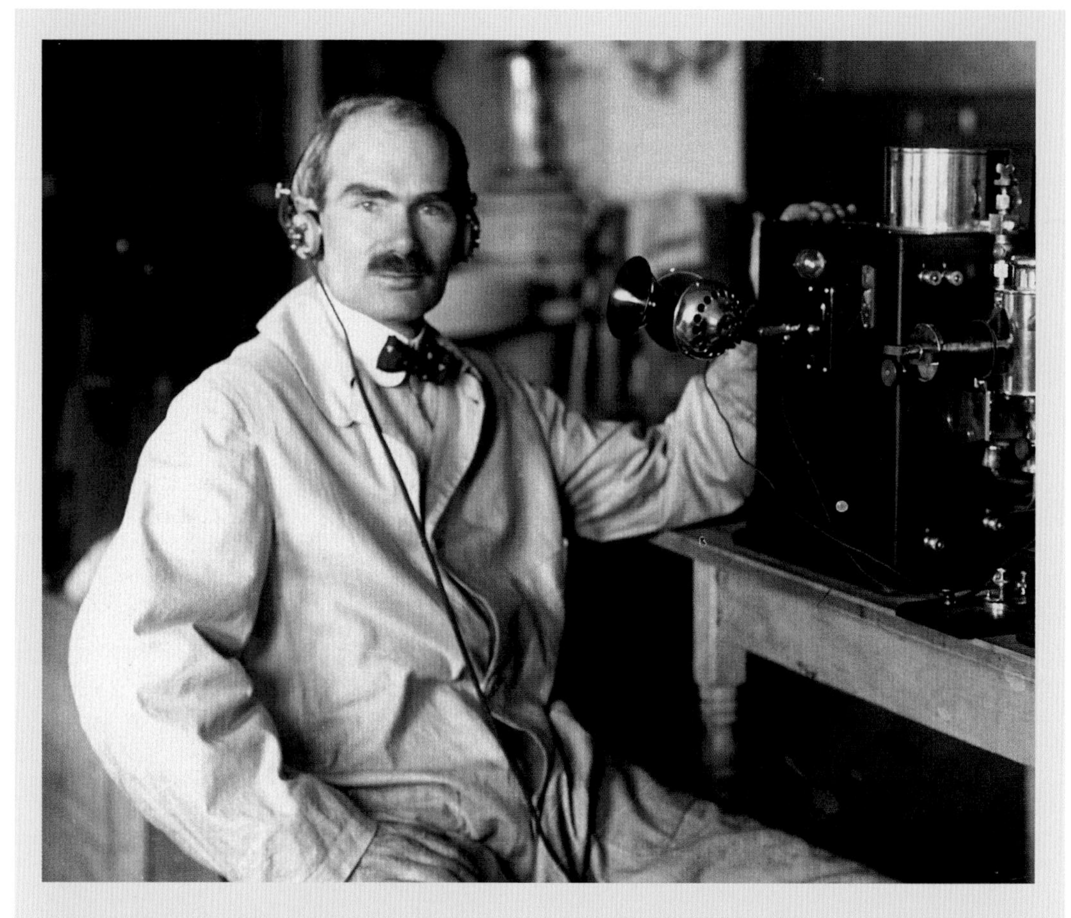

The man who is generally regarded as the first to transmit entertainment by radio is Lee De Forest (1873–1961), who liked to refer to himself as the "father of radio". Born into modest circumstances in Iowa, De Forest attempted unsuccessfully to gain employment with both Tesla and Marconi. At the turn of the twentieth century he moved to New York, where a shady financier helped him form his first company. The latter's aggressive promotion of the firm helped create the impression that a financial bubble was forming around radio, which may have helped to spook Tesla investors such as Morgan.

In 1906 De Forest invented a vacuum tube that he dubbed the "Audion", the first effective device for amplifying radio signals. To the consternation of men such as Tesla, De Forest's understanding of how the device worked seemed quite limited, but it did make radio broadcasting feasible and established the vacuum tube as the foundational component of electronics. De Forest eventually earned over 180 patents and started a number of companies, boasting that he had made and lost multiple fortunes over the course of his life. Like Tesla, he died a poor man, though unlike Tesla he married four times.

ABOVE. Photograph of radio pioneer Lee De Forest from the early 1900s.

EDWIN ARMSTRONG

Edwin Armstrong (1890–1954) had been interested in electrical devices since boyhood, constructing a backyard antenna that he delighted in climbing. A graduate of Columbia University, in 1913 Armstrong developed what he called his "regenerative circuit", which could amplify the strength of radio signals by hundreds of times. After serving in World War I, Armstrong retained legal counsel to fight the claims of De Forest and others that they had discovered regeneration first. Armstrong sold the commercial rights to this work to the Westinghouse company, but De Forest eventually prevailed in a patent lawsuit.

While serving in the radio corps during World War I, Armstrong developed his "superheterodyne" radio receiver, which was both more sensitive and more specific in tuning into radio signals. Armstrong interested David Sarnoff (see p.117) and RCA in his design, and beginning in 1924, his "Radiola" sets were a huge commercial success. In 1922, he developed a technique of "super-regeneration", which was even more effective, selling his patent to RCA and becoming the company's largest shareholder. Armstrong's other great invention was FM (frequency modulation), which eliminated most electrical interference.

Armstrong had a quarrel with RCA over his FM patents. Initially, the company was reluctant to adopt FM, because it might undermine investments in AM (amplitude modulation) broadcasting. Eventually, RCA came around, offering Armstrong $1 million for the right to use FM, but he refused out of fairness to other companies. Over the years, Armstrong's fortune began to be depleted by legal disputes, and in 1954 these frustrations led him to strike his wife with a fireplace poker. Distraught, he jumped to his death from the window of his 13th-floor apartment.

RIGHT. Edwin Armstrong, discoverer of the superheterodyne receiver.

Brig. General David Sarnoff, Chairman of the Board, Radio Corporation of America

Sees No. 1 wish come true!

Television Tape Recording by RCA Opens New Era of Electronic Photography

In 1956, RCA's General Sarnoff will celebrate his 50th year in the field of radio. Looking ahead to that occasion, three years ago, he asked his family of scientists and researchers for three gifts to mark that anniversary: (1) A television tape recorder, (2) An electronic air conditioner, (3) A true amplifier of light.

Gift No. 1—the video tape recorder—has already been successfully demonstrated, two years ahead of time! Both color and black-and-white TV pictures were instantly recorded on magnetic tape without any photographic development or processing.

You can imagine the future importance of this development to television broadcasting, to motion pictures, education, industry and national defense. And you can see its entertainment value to you, in your own home. There the tape equipment could be used for home movies, and—by connecting it to your television set—you could make personal recordings of your favorite TV programs.

Expressing his gratitude for this "gift," General Sarnoff said it was only a matter of time, perhaps two years, before the finishing touches would bring this recording system to commercial reality. He described this RCA achievement as the first major step into an era of "electronic photography."

Such achievements as this, stemming from continuous pioneering in research and engineering, make "RCA" an emblem of quality, dependability and progress.

RADIO CORPORATION OF AMERICA

World leader in radio — first in television

14

TIME, FEBRUARY 15, 1954

DAVID SARNOFF

Unlike Fessenden, De Forest and Armstrong, David Sarnoff (1891–1971) was not primarily an inventor but a businessman. Born in Russia, Sarnoff grew up in straitened circumstances in New York City. Sarnoff went to work for Marconi in 1906, soon becoming a manager. One of the keys to Sarnoff's success was his insight that radio could be used not only for point-to-point transmission but also for broadcasting, with many receivers tuning in to the same transmission. When American Marconi was purchased and turned into RCA, Sarnoff advanced this proposal, but it fell on deaf ears.

To prove his point, Sarnoff arranged to broadcast a heavyweight boxing match, attracting a huge audience, and soon the potential of broadcasting became apparent to all. RCA formed NBC, the first radio network in the US. Sarnoff became RCA's president in 1930, and soon committed the company to the development of television, offering the first regularly scheduled broadcasts in 1939. In the 1950s, Sarnoff led RCA into colour television. Thanks in large part to Sarnoff's vision, RCA became one of the most important electronics firms of the twentieth century.

The stories of Fessenden, De Forest, Armstrong and Sarnoff leave one of Tesla's most important questions unaddressed – namely, how could radio signals be transmitted over long distances, beyond the line of sight? Low-frequency AM signals can travel as ground waves, following the earth's contour. Moreover, in the short-waveband range, signals can be reflected back to earth by a layer of the atmosphere known as the ionosphere, enabling them to travel around the world. Tesla was at least partially right, however. At frequencies above 30 megahertz (MHz) or so, radio waves do tend to be limited to line of sight.

Tesla was right about another thing, too. At extremely low frequencies, the earth can serve as a conductor of radio signals, a potential that the world's major military powers exploited for decades in communicating with submarines. Ordinary radio waves do not penetrate far below the ocean, but at very low frequencies (corresponding to wavelengths of thousands of kilometres), signals can penetrate deep into the ocean. On the downside, such communication was one-way only, and information could be transmitted only at a very low rate of a few characters per minute.

OPPOSITE. 1954 *Time* article on RCA, featuring David Sarnoff with the company's first videotape recorder.

BELOW. The third US radio conference, 1924. Second from left is David Sarnoff, and fourth from left is future US President Herbert Hoover.

After Wardenclyffe

Unable to attract the funds to make Wardenclyffe operational, Tesla seems to have fallen into a state of depression in 1906. His plaintive letters to Morgan ceased. He stopped tending to his personal appearance and hygiene. He did not socialize for months. Only in 1907 did his spirits begin to revive, in part because his election to the New York Academy of Sciences led him to believe that his reputation was not entirely lost. Soon he was back in action, mortgaging his properties and soliciting loans from his associates.

Tesla began writing again, publishing essays in a wide variety of newspapers and journals on topics as varied as interplanetary communications, his worldwide wireless system, a radio-controlled torpedo, the effects of transcranial currents (which Tesla had employed on himself during his depression), his complaints against Marconi and others for pirating his wireless patents, and how explosives could be used to generate tsunamis that would wipe out approaching enemy flotillas. Tesla also envisioned an aircraft that would be held aloft by wireless energy transmissions from the ground.

Perhaps sensing that his wireless project was going nowhere, in 1909 Tesla returned to a dream of his childhood: the development of turbines. In particular, he now focused on the development of a bladeless turbine. His idea was rooted in viscosity, the internal friction between molecules in a flowing liquid, which could be used to turn a series of closely spaced discs. Tesla reported that his design could deliver several dozen times more horsepower per unit of engine weight than other engines, making automobiles and aeroplanes a great deal more efficient.

An opportunity for Tesla to press his case against Marconi as the true father of radio arose in 1912, when the German company Telefunken mounted an attempt to overturn Marconi's patents. Tesla was placed on a retainer of $1,000 per month, not only for help in mounting a legal case but also for his technical expertise. Thanks in part to Tesla's contributions, the output of one of the company's stations at Sayville, New York was tripled in only a few months, positioning it to become the US's most important transmitter during World War I.

OPPOSITE LEFT. 1913 illustration of one of Tesla's bladeless turbines, developed from 1906.

OPPOSITE RIGHT. A small model of the bladeless turbine.

THE FATHER OF RADIO CONTROL

In 1911, an inventor and entrepreneur named John Hays "Jack" Hammond, Jr. (1888–1965) reached out to Tesla with a proposal to form the Tesla-Hammond Wireless Development Company, which would focus first on developing Tesla's radio-controlled torpedo. Tesla had augmented the security of the system some years earlier by requiring a combination of radio frequencies to be transmitted for it to respond, thereby making it unlikely that an enemy could seize control of it. Patented in 1903, this system would be as secure as any combination lock.

Yet Tesla declined to collaborate, preferring to focus on his bladeless turbine and the dream of reviving his worldwide wireless system. Hammond proceeded with development and eventually received large payments from RCA and the military for his work, becoming known as the "father of radio control". He laid the groundwork for the development of missile guidance systems, unmanned aircraft, and today's aerial drones. Had Tesla partnered with Hammond, he would probably have made a fortune, but because his patents expired, he ended up receiving nothing.

ABOVE. John Hays "Jack" Hammond, Jr., the "father of radio control", as photographed in 1922.

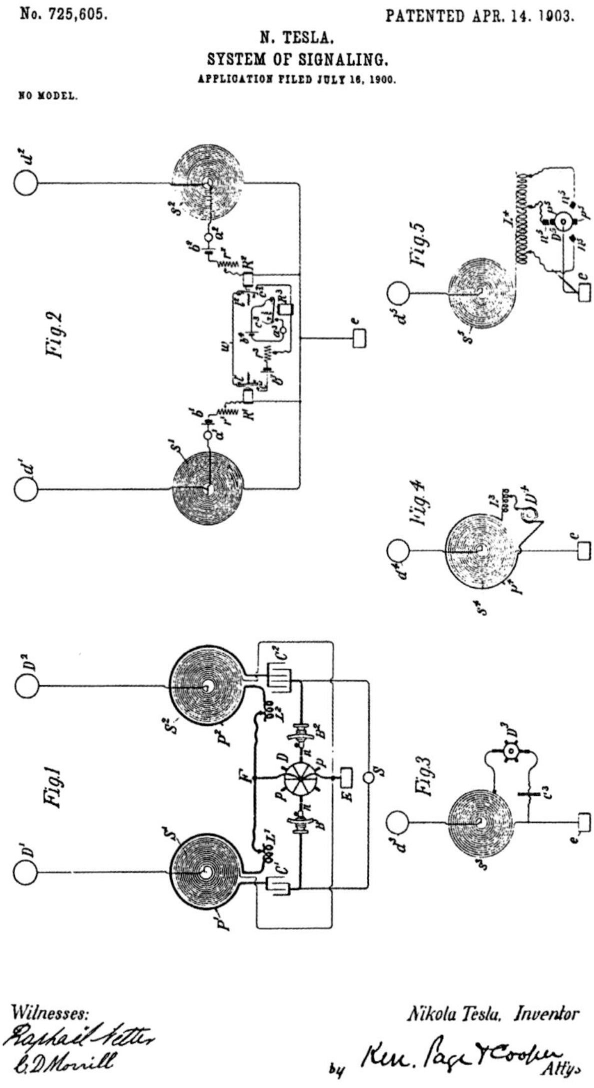

In the courtroom where the Marconi case was heard, many of the greatest inventors of the age testified, including Tesla. One of these men, John Stone, was one of the pioneers of radio tuning, who held 120 patents and received medals from the Franklin Institute and the Institute of Radio Engineers. Having reviewed the history of radio, Stone testified that he was surprised when his research disclosed the true magnitude of Tesla's contributions.

> *I think we all misunderstood Tesla. He was so far ahead of his time that the best of us mistook him for a dreamer.*

Telefunken's case was rendered moot when the US entered World War I against Germany in 1917, but the legal wrangling over priority for radio's development continued into the 1940s.

Tesla had another idea of interest to the US military during World War I – namely, RADAR. German U-boats were sinking allied ships at an alarming rate, and Tesla published the outlines of a possible solution in 1917:

> *If we can shoot out a concentrated ray comprising a stream of minute electric charges vibrating at tremendous frequency, and then intercept this ray, after it has been reflected by a submarine hull, and cause this intercepted ray to illuminate a fluorescent screen on the same or another ship, then our problem of locating the hidden submarine will have been solved.*

While the attenuation of electromagnetic waves in sea water would render this vision impractical, Tesla's idea ultimately proved useful for monitoring aircraft and weather systems in the atmosphere.

World War I was not a happy time for Tesla. He was warned by an advisor that his unpaid debts could lead to public disclosure of his straitened finances, and the prophecy came true when Tesla was taken to court for his failure to pay $935 in personal taxes. A March 1916 article (see opposite for full piece) in the *New York Tribune* trumpeted the news:

> *Thirty-five years of unceasing efforts for humanity by Nikola Tesla, inventor and engineer, whose name is known wherever electricity is used, have proved to be a labor of love alone. The man who ranks with Edison and Marconi yesterday confessed himself not only unable to pay his personal taxes, but dependent on credit for the necessities of life.*

No. 725,605. PATENTED APR. 14. 1903.

N. TESLA.

SYSTEM OF SIGNALING.

APPLICATION FILED JULY 16, 1900.

NO MODEL.

Witnesses: Nikola Tesla, Inventor

The New York Tribune March 18, 1916

TESLA, INVENTOR, CAN'T PAY TAXES

Once Owner of $500,000 Plant Sued for $935—Now Lives on Credit.

Thirty-three years of unceasing efforts for humanity by Nikola Tesla, inventor and engineer, whose name is known wherever electricity is used, have proved to be a labor of love alone. The man who ranks with Edison and Marconi yesterday confessed himself not only unable to pay his personal taxes, but dependent on credit for the necessities of life.

As a living proof of the inability of a genius to convert his gifts into cash, Tesla presented himself as Exhibit A in supplementary proceedings instituted by Corporation Counsel Hardy for $935 personal taxes. Justice Finch, of the Supreme Court, has appointed R. McMath receiver for the inventor. Eighty independent inventions, every one of which has been adopted by the engineering world, have failed to place a penny in the pocket of the man who created them.

Only a few years ago Tesla owned nine-tenths of the stock of the Nikola Tesla Company, a $500,000 corporation, with offices at 8 West Fortieth Street. All of this stock was used as security on loans to be used in making new inventions. Just now the inventor has neither place of business nor home. He is occupying a room at the Waldorf-Astoria which, according to his testimony, he is allowed to occupy while owing the hotel a bill that has been unpaid for years.

"How do you live?" asked Corporation Counsel Hardy, who examined Tesla.

"Why, mostly on credit," he answered, after some hesitation.

"Has the Tesla company any assets?"

"No, not now," said its founder. "It is getting a little in royalties, but not enough to pay expenses."

"Do you own any patents?"

"No, not any more. All of them have been assigned to various creditors or to the company. There were more than 200 patents."

In answer to a question whether there were any judgments against him Tesla said there were scores of them—and more pending. He said he had no money in the bank nor elsewhere. When asked whether he had any jewelry he said he abhorred it.

Probably the greatest of Tesla's numerous inventions was that of the alternating motor, into which was compressed the maximum of horse power in the minimum of space and weight of material. His 200-horsepower engine was of such size that it could be placed in a receptacle little larger than a hat box.

A 10-horsepower engine was constructed by Tesla of such diminutive proportions that it might be dangled from the hand by a string. The possibilities of this method of storing power, as well as the numerous uses to which an alternating motor might be put, were such that scientists unhesitatingly said it would bring untold wealth to its inventor.

The Tesla arc light, methods of eliminating waste in distribution of electricity, development of wireless telegraphy and numerous other practical inventions at various times promised to bring wealth to their creator. But money was as hard to lure into the pockets of Tesla as engineering ideas were easy to bring from his brain. Practically all of his engineering dreams have come true without bringing about the realization of one of his modest dreams of financial independence.

Largely as a matter of form, the Justice appointed a receiver for Tesla because of the city's claim against him. Neither Tesla nor the receiver expects to find anything to receive.

ABOVE. *New York Tribune*, 18 March 1916; article detailing Tesla's desperate finances.

OPPOSITE. Diagram from Tesla's 1903 patent application for a system of wireless signalling.

BITTERNESS

Impoverished, marginalized and even ignored, Tesla became increasingly frank in expressing his displeasure at the successes of his rivals. Regarding Marconi, for example, Tesla believed that anyone with any degree of familiarity with the technology could not fail to recognize that Marconi's work rested on Tesla's foundation. Yet a 1904 US Patent Office ruling gave the patent for radio to Marconi, and five years later, Marconi shared the Nobel Prize with Karl Braun. Years later, Tesla is reputed to have said of Marconi, "He is a donkey."

While it might appear that Tesla simply meant to insult Marconi by comparing him to an animal often held in low regard, there may have been more to it than that. Donkeys, of course, are often used as pack animals, and in Tesla's mind, Marconi had done little more than stubbornly refine the apparatus necessary to transmit radio signals over progressively greater distances. The breakthrough, thought Tesla, was his, and honouring Marconi made a mockery of the whole endeavour.

Tesla's bitterness was undoubtedly enhanced by the belief that, had he so chosen, he could have transmitted transatlantic signals before Marconi. Instead, however, he had been pursuing bigger game – the global wireless transmission of power. From the standpoint of an investor such as Morgan, however, it may have appeared differently. Namely, while Tesla could talk a good game, it was Marconi who actually came through with a workable device, and at a substantially lower cost.

ABOVE. Wireless Marconi receiver, with which the Italian inventor received transatlantic wireless signals for the first time in December 1901, helping to bring the new technology of radio within the reach of ordinary people.

OPPOSITE. Marconi's experimental tuned transmitter.

Throughout much of his later life, Tesla seems to have been tortured by a sense that lesser men were surpassing him and gloating over their triumph. Yet Tesla would not give them the satisfaction of admitting defeat. As he wrote in *My Inventions*:

> *I am unwilling to accord to some small-minded and jealous individuals the satisfaction of having thwarted my efforts. These men are to me nothing more than microbes of a nasty disease. My project was retarded by laws of nature. The world was not prepared for it. It was too far ahead of its time. But the same laws will prevail in the end and make it a triumphal success.*

While refusing to grant that he was in the least affected by the injustice the world has done him by elevating his rivals, Tesla appears to betray an awareness of its toxic effects. In saying that such men are to him nothing more than "microbes of a nasty disease", he implies that such resentments can infect and poison a person. Though unable to reconcile himself to the situation, he may have been sufficiently self-aware to recognize the toll his bitterness was taking on him.

Although Tesla said more than once that it is above all in adversity that a man's strength reveals itself, he himself seems not to have fared well under such circumstances. He continued to dream, to work and to invent, but he also began spending more of his time writing self-aggrandizing articles and letters to popular periodicals. He guarded jealously the credit he believed rightly belonged to him and became increasingly envious of others who displaced him on the world's stage.

His dreams had always been difficult for him to distinguish from reality, but they now seemed so undeniable to him that he would often share new insights with reporters and the public almost as soon as they occurred to him, not taking the time to think them through or put them to the test. In other words, he spent less and less time critically examining and verifying his own assumptions. Some of his associates began to fear that, in terms of his credibility, he was becoming his own worst enemy.

News of Tesla's financial difficulties certainly did not help his case. While rivals such as Edison and Marconi had built thriving businesses that bore their names, Tesla's firms had largely foundered, a desperate situation exacerbated by the expiration of his patents. When he criticized great inventors, titans of industry and epochal scientists, many observers saw at best only sour grapes and at worst the ruminations of a mad scientist, a genius who appeared to be taking leave of his senses.

ABOVE. Marconi and his assistants hoist a radio aerial at Signal Hill in Newfoundland, in preparation for receiving the first ever transatlantic radio signals, 1901.

EINSTEIN

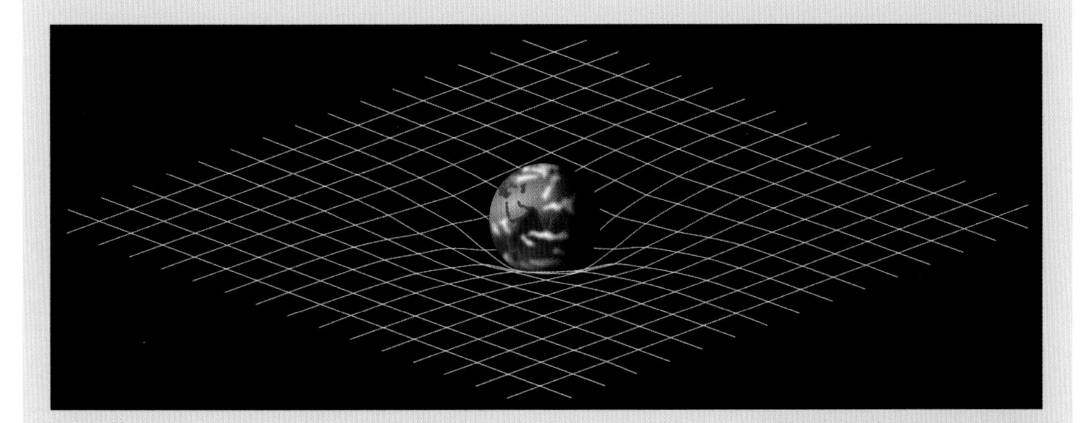

The full depth of Tesla's capacity for resentment is revealed by his attitude towards Albert Einstein, arguably the most important scientist of the twentieth century, who received the Nobel Prize for Physics in 1921. Tesla disagreed with Einstein on a number of points. In sharp contrast to Einstein, he believed that some forms of energy could travel faster than light, and he regarded the notion that space could be distorted by gravity as absurd. On the latter point, he wrote:

> *I hold that space cannot be curved, for the simple reason that it can have no properties. ... Of properties we can only speak when dealing with matter filling the space. To say that in the presence of large bodies space becomes curved is equivalent to stating that something can act upon nothing. I refuse to subscribe to such a view.*

In contrast to Marconi, whom Tesla resents for receiving credit for work that he performed, Tesla feels aggrieved at Einstein for promoting views fundamentally at odds with his own sense of how the universe works. If Einstein were right, then the Newtonian model of physics would be in crucial respects wrong, and this was a notion that Tesla – in many respects an electrical and mechanical engineer whose work was quite at home in a Newtonian universe – could not abide.

Tesla seems to have regarded pure theoreticians such as Einstein as living in a shadow world of blackboards and equations, while he rolled up his sleeves, experimented, and produced devices that actually worked in the real world. For example, he described the theory of relativity as nothing more than

> *a mass of error and deceptive ideas violently opposed to the teachings of great men of science of the past and even to common sense. The theory wraps all these errors and fallacies and clothes them in magnificent mathematical garb which fascinates, dazzles, and makes people blind to the underlying error. The theory is like a beggar clothed in purples whom ignorant people take for a king. Its exponents are brilliant men, but they are metaphysicists rather than scientists.*

ABOVE. Schematic diagram depicting the bending of space by the gravitational force of a massive object, a concept that Tesla ridiculed.

RIGHT. Albert Einstein in 1921, the year he received the Nobel Prize for Physics.

On 7 November 1915, just a week before the official announcement of the year's Nobel Prizes, a front-page story announced that Nikola Tesla was to share the award with Thomas Edison:

> *Nikola Tesla, who with Thomas Edison is to share the Nobel Prize in Physics, according to a dispatch from London, said last evening that he had not yet been officially notified of the honor. His only information on the matter was the dispatch in the New York Times.*

> *"I have concluded," he said, "that the honor has been conferred upon me in acknowledgement of a discovery announced a short time ago which concerns the transmission of electrical energy without wires. This discovery means that electrical effects of unlimited intensity and power can be produced, so that not only can energy be transmitted for all practical purposes to any terrestrial distance, but even effects of cosmic magnitude may be created. ..."*

Mr. Tesla refused to go further into the matter. He said he thought Mr. Edison was worthy of a dozen Nobel prizes. He knew nothing of the discovery, he said, that induced the authorities in Sweden to confer the great honor on Mr. Edison.

While undoubtedly aware that receiving the Nobel Prize could both supply needed funds and do much to rehabilitate his standing in the eyes of the public, Tesla also could not resist the impulse to put the prize in what he regarded as its proper place, in relation to his own legacy, writing to friends:

> *I have not less than four dozen of my creations identified with my name in technical literature. These are honors real and permanent, which are bestowed,*

The New York Times May 19, 1917

Edison Medal Goes to Nikola Tesla

Nikola Tesla, inventor, was last night adjudged by the American Institute of Electrical Engineers to have contributed the greatest progress to electrical science and electrical engineering during the year 1916. The Edison medal which is awarded each year to the person deemed foremost in this field was presented to him at the annual meeting of the institute, in the Engineering Societies Building, No. 33 West Thirty-ninth street.

The medal is a gold plaque, with the bust of Mr. Edison raised upon one surface. It was presented by H. W. Buck, president of the institute.

The annual report of the Board of Directors was presented, showing a total membership of 8,710. Total revenue during the year was $117,843.68, and total expenses $106,069.21.

OPPOSITE. *New York Times*, 6 November 1915; dispatch announcing that Tesla will share the Nobel Prize in Physics with Thomas Edison.

RIGHT. Lawrence Bragg, who at age 25 shared the Nobel Prize in Physics with his father for their development of X-ray crystallography. Bragg later nominated Watson, Crick and Wilkins for the Nobel Prize for their demonstration of the double-helix structure of DNA.

BELOW. The Edison Medal, whose recipients included Tesla.

not by a few who are apt to error, but by the whole world which seldom makes a mistake, and for any of these I would give all the Nobel prizes during the next thousand years.

However, when the official announcement of the Nobel Prize in Physics came from Stockholm on 14 November, neither Edison nor Tesla was so honoured. Instead the Prize went to the father-son team of William H. and W. Lawrence Bragg – making the latter, at the age of 25 years, the youngest science winner in history – for their work developing X-ray crystallography, which would later play a crucial role in illuminating the double-helix structure of DNA.

It is not clear why Edison and Tesla were not honoured. Some speculated that Tesla had declined the prize, refusing to share it with Edison. Others suggested that it was Edison who declined the award, in part because the wealthy and famous man knew that by doing so he would deprive Tesla of needed funds. However, there is no hard evidence for either view, and the Nobel Foundation simply explained that no prospective recipient would ever be denied the prize for having refused it in advance.

A year later, Tesla was informed by the American Institute of Electrical Engineers that he had been selected to receive the 1916 Edison Medal. The award had been established in 1904 by a group of Edison's

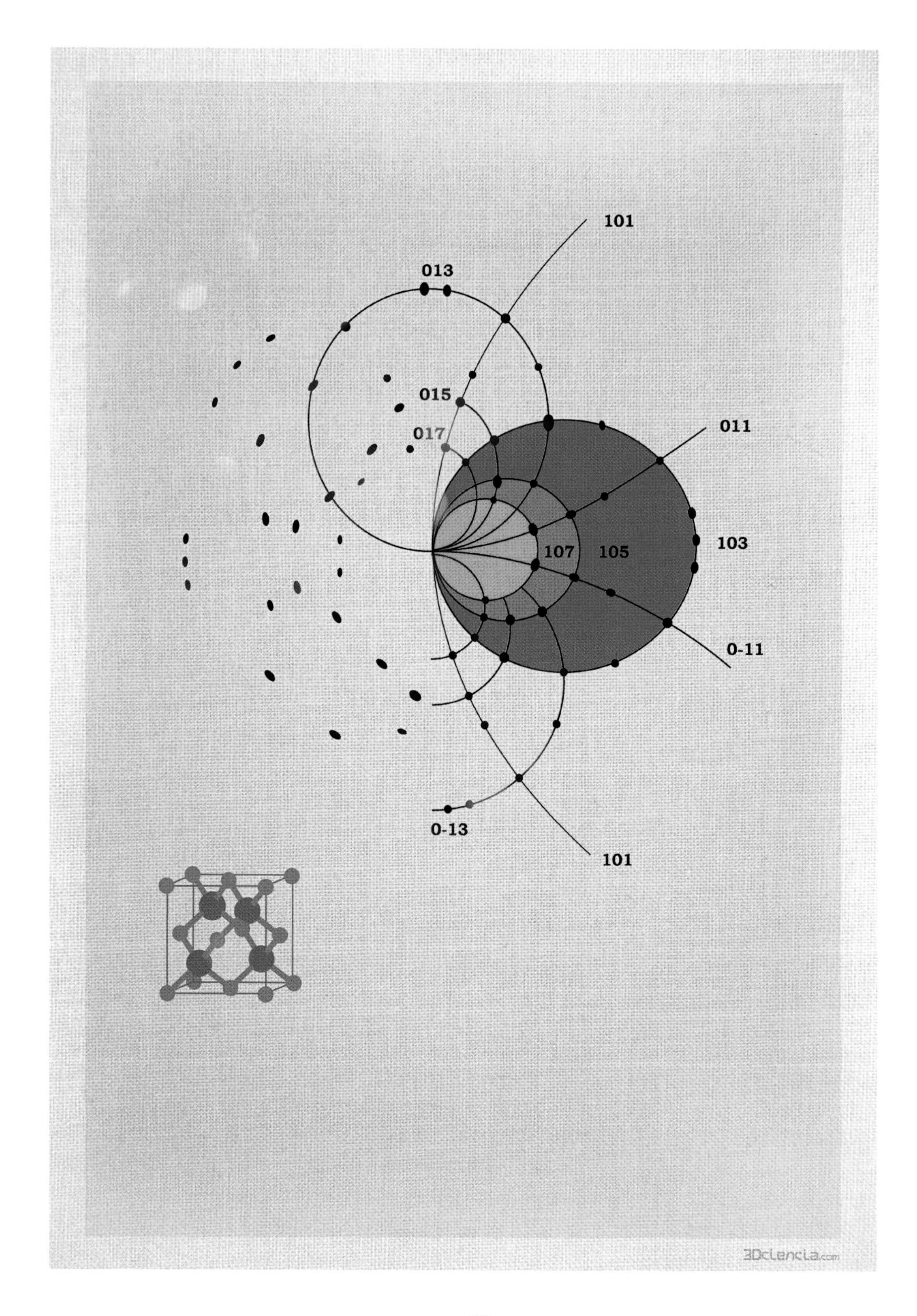
101
013
015
017
011
107
105
103
0-11
0-13
101
3Dciencia.com

friends and admirers to honour a living electrician for "meritorious achievement in electrical science and art". Prior recipients included Elihu Thomson (inventor and founder of major electric companies in Europe and the US); Westinghouse; and Alexander Graham Bell, inventor of the telephone.

Tesla's high regard for his own importance made it difficult for him to respond to such awards in a graceful manner:

> *You propose to honor me with a medal which I could pin upon my coat and strut for a vain hour before the members and guests of your institute. You would bestow an outward semblance of honoring me but you would decorate my body and continue to let starve, for failure to supply recognition, my mind and its creative products which have supplied the foundation upon which the major portion of your institute exists.*

The idea that he, Tesla, should receive a medal bearing Edison's name struck him as equally out of proportion:

> *And when you would go through the vacuous pantomime of honoring Tesla you would not be honoring Tesla but Edison, who has previously shared unearned glory from every previous recipient of the award.*

Eventually, however, Tesla agreed to accept the award, perhaps because he had learned that Edison planned not to attend the ceremony, and he dutifully showed up at the appointed hour for the evening's festivities. Yet as the time for making the presentation approached, the institute representative who nominated Tesla realized that he was nowhere to be found. Frantic, he left the building and ventured to Bryant Park, where he found Tesla calmly feeding the pigeons.

With Tesla now back in attendance, the presentation of the award closed with the following lines, paraphrasing words Alexander Pope had used to describe Isaac Newton:

> *Nature and nature's laws lay hid in night;*
> *God said, let there be Tesla, and all was light.*

Tesla rose and even managed to express appreciation for Edison, remarking that despite a lack of theoretical training, his work ethic and perseverance had enabled him to achieve notable results.

Continuing with his acceptance speech, Tesla described his method of invention, which he contrasted with that of Edison. While Edison would experiment tirelessly with progressive iterations of an idea, in Tesla's case,

> *Nature has given me a vivid imagination which,*

ABOVE. Alexander Graham Bell, inventor of the telephone, c.1916.

OPPOSITE. Illustration of an X-ray diffraction pattern, indicating the structure of the crystal located in the lower left of the image.

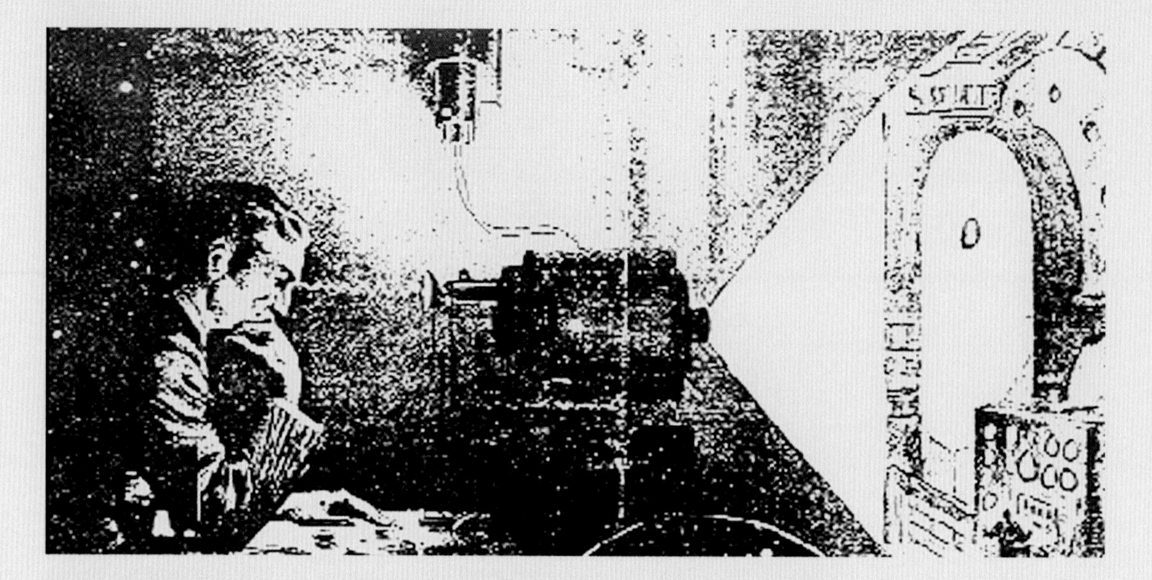

through incessant exercise and training, study of scientific subjects and verification of theories through experiment, has become very precise, so that I have been able to dispense, to a large extent, with the slow, laborious, wasteful and expensive process of practical developments of the ideas I conceive. It has made it possible for me to explore extended fields with great rapidity and get results with the least expenditure of vital energy.

Tesla then outlines what might be called his philosophy of work and life, which helps to explain what audience members may have read about the state of his affairs in the press:

By this means I have it in my power to picture the objects of my desires in forms real and tangible and so rid myself of that morbid craving for perishable possessions to which so many succumb. I may say, also, that I am deeply religious at heart, although not in the orthodox meaning, and that I give myself to the constant enjoyment of believing that the greatest mysteries of our being are still to be fathomed and that, all the evidence of the senses and the teachings of exact and dry sciences to the contrary notwithstanding, death itself may not be the termination of the wonderful metamorphosis we witness. In this way I have managed to maintain an undisturbed peace of mind, to make myself proof against adversity, and to achieve contentment and happiness to a point of extracting some satisfaction even from the darker side of life, the trials and tribulations of existence. I have fame and untold wealth, more than this, and yet how many articles have been written in which I was declared to be an impractical unsuccessful man, and how many poor, struggling writers have called me a visionary. Such is the folly and shortsightedness of the world!

Tesla then goes on to describe both his presence at the ceremony and the continuation of his very existence in terms of the miraculous, singling himself out among the other men who have previously received the award:

Of course, the men who have received this medal have fully deserved it, in that respect, because they were alive when it was conferred upon them, but none has deserved it in anything like the measure I do, when it comes to that feature. In my youth my ignorance and lightheartedness brought me into innumerable difficulties, dangers and scrapes, from which I extricated myself as by enchantment. That occasioned my parents great concern more, perhaps, because I was the last male than because I was of their own flesh and blood. You should know that Serbians desperately cling to the preservation of the race. I was nearly drowned a dozen times. I was

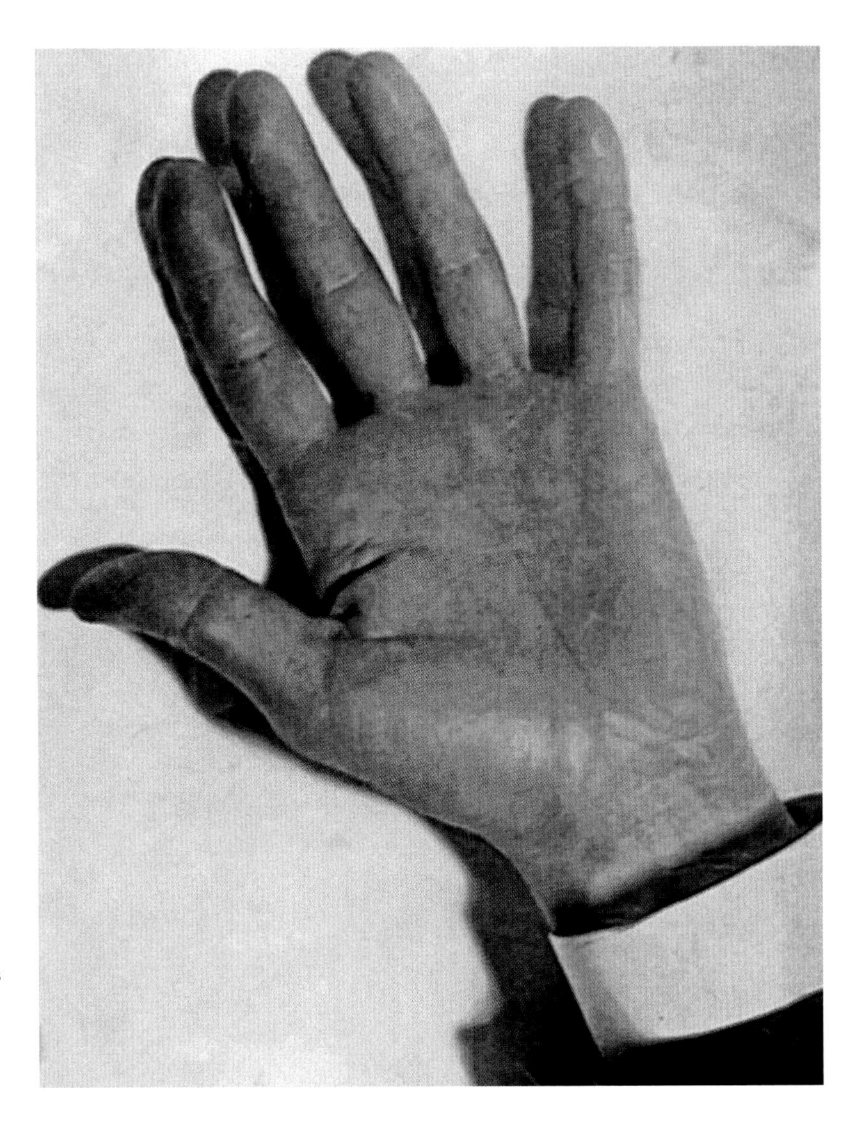

OPPOSITE. Artist's depiction of Tesla's "thought camera", which would project what a person was thinking onto a screen.

RIGHT. Nikola Tesla's hand, illuminated by his artificial daylight invention.

almost cremated three or four times and just missed being boiled alive. I was buried, abandoned and frozen. I have had narrow escapes from mad dogs, hogs and other wild animals. I have passed through dreadful diseases. I have been given up by physicians three or four times in my life for good. I have met with all sorts of odd accidents. I cannot think of anything that did not happen to me, and to realize that I am here this evening, hale and hearty, young in mind and body, with all these fruitful years behind me, is little short of a miracle.

As these passages from Tesla's speech indicate, any pretence of modesty on Tesla's part was just that – a pretence. In his own mind, the singularity and significance of his life and work could hardly be exaggerated. Far from being undermined, his sense of his own importance was magnified by the many adversities he had endured. Even now he was preparing to rise again from the ashes, promoting such revolutionary ideas as a means to illuminate the sky to avoid collisions between seafaring vessels and a device for projecting a person's thoughts onto a screen.

Birthday Press Conferences

In July 1931, Tesla's portrait appeared on the cover of *Time*, in celebration of the inventor's 75th birthday. The accompanying story reminds readers that they can no longer see Alessandro Volta, André Ampère, Georg Ohm, Charles de Coulomb, Luigi Galvani or James Watt – all scientists and inventors who contributed to our understanding of electricity – but they can still see Nikola Tesla, a "tall, meagre, eagle-headed man".

The reporter finds Tesla, "not without some difficulty", not at work in his laboratory with electrical discharges streaming around him, but in seclusion on the 20th floor of Manhattan's Governor Clinton Hotel. Tesla appears "pale but healthy, thin to ghostliness but strong and alert as ever", only the sparkle of his blue eyes and the shrillness of his voice indicating his "psychic tension". It becomes almost immediately apparent that Tesla continues to dream.

"To Tesla," says the article, "all the world's a powerhouse". Thanks to the earth's electrical resonance, he believes, all that is required to provide unlimited power, under no one's financial control, to anyone on earth is to generate electricity in tune with the earth. For years, Tesla's annual interviews have been a "rehash" of his vision of broadcasted power, but this year he reveals that he is at work on two new ideas.

First, he is perfecting a purely mathematical explanation of phenomena that Einstein has described, but Tesla's account will "tend to disprove the Einstein theory". He continues, "When I am ready to make a full announcement it will be seen that I have proved my conclusions". Second, he is developing a "new source of power", one to which "no previous scientist has turned". Not only will it provide a new source of power, but it will also "throw light on many puzzling phenomena of the cosmos".

Tesla goes on to say that his new power source has nothing to do with so-called atomic energy. For one thing, "there is no such energy in the sense usually meant". Tesla reports that he has been splitting atoms and found that no energy is released. Those assembled at the interview naturally badgered Tesla to reveal the nature of his new energy source, but he would only say that it will "make it possible for man to transmit energy in large amounts from one planet to another, absolutely regardless of distance".

Concerning transmissions between planets, Tesla states:

I think that nothing can be more important than interplanetary communication. It will certainly come someday, and the certitude that there are other

OPPOSITE. Tesla on the cover of the 20 July 1931 issue of *Time* magazine.

human beings in the universe, working, suffering, struggling, like ourselves, will produce a magic effect on mankind and will form the foundation for a universal brotherhood that will last as long as humanity itself.

The story lists the many eminent scientists, engineers and titans of industry who have sent congratulatory messages to Tesla marking his birthday. Some express the hope that he still has in him "one more astounding new device for mankind". Yet, the reporter concludes, "It is improbable that he will ever design such a device on paper, let alone in a machine shop, although before his mind's eye he may see it in every detail, motion, and defect. He is a great visualizer."

The sense of scepticism implicit in the description of Tesla as a "great visualizer" would be amplified at subsequent Tesla birthday press conferences. For example, in 1934 Tesla announced that he was hard at work on a new particle beam weapon that would "bring down a fleet of 10,000 enemy airplanes at a distance of 250 miles from a defending nation's

LEFT. Press photo of Tesla from his 1935 birthday press conference

BELOW. Letter from William Lee, President of the American Institute of Electrical Engineers, congratulating Tesla on his 75th birthday.

border and cause armies of millions to drop dead in their tracks". He would later dub a version of this new weapon "Teleforce".

In 1938, the *New York Times* reported that Tesla had postponed his 82nd birthday press conference so that "he could complete a mechanical therapeutic device" and prepare to give a demonstration to the press "in two weeks". Tesla stated that he had perfected the device years ago before his laboratory burned down, that he had used it to the advantage of his friend, Mark Twain, and he need merely place "one of these devices in every newspaper office and the proverbial 'editorial grouch' will be gone".

Tesla also reported that he had perfected an invention that would benefit the poultry industry by improving "the size, quality, and quantity of eggs and chickens". Tesla is described as living on the 33rd floor of the hotel New Yorker, where, "in one large room crowded with plans, boxes, and technical references, he conducts his experiments and research". While duly reporting Tesla's announcements, the story also notes that some scientists have challenged his claims as "fantastic".

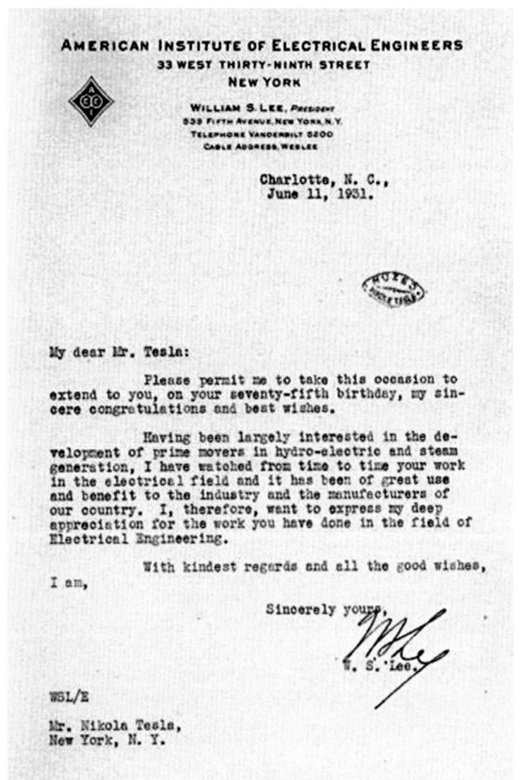

AMERICAN INSTITUTE OF ELECTRICAL ENGINEERS
33 WEST THIRTY-NINTH STREET
NEW YORK

WILLIAM S. LEE, PRESIDENT
539 FIFTH AVENUE, NEW YORK, N.Y.
TELEPHONE VANDERBILT 5200
CABLE ADDRESS, WESLEE

Charlotte, N. C.,
June 11, 1931.

My dear Mr. Tesla:

Please permit me to take this occasion to extend to you, on your seventy-fifth birthday, my sincere congratulations and best wishes.

Having been largely interested in the development of prime movers in hydro-electric and steam generation, I have watched from time to time your work in the electrical field and it has been of great use and benefit to the industry and the manufacturers of our country. I, therefore, want to express my deep appreciation for the work you have done in the field of Electrical Engineering.

With kindest regards and all the good wishes, I am,

Sincerely yours,

W. S. Lee.

WSL/E

Mr. Nikola Tesla,
New York, N. Y.

THE COSMIC RAY

In 1937, the *New York Times* Tesla birthday story included reports of discoveries that would make it possible "to communicate with planets and produce radium in unlimited quantity for $1 per pound". Although Tesla displayed no apparatus or sketches to illustrate his claims, he did state that he would be able to give a demonstration in "only a little time". Describing his work on cosmic rays, Tesla stated:

> *I am proud of these discoveries, because many have denied that I am the original discoverer of the cosmic ray. I was fifteen years ahead of the other fellows, who were asleep. Now no one can take away from me the credit of being the first discoverer of the cosmic ray on earth.*

Describing his efforts to communicate with other worlds, Tesla continued:

> *I am expecting to put before the Institute of France an accurate description of the devices with data and calculations and claim the Pierre Guzman prize of 100,000 francs for means of communication with other worlds, feeling perfectly sure that it will be awarded to me. The money, of course, is a trifling consideration, but for the great historical honour of being the first to achieve this miracle I would be almost willing to give my life.*

Concerning another invention, his "atom-smashing tube", Tesla expressed annoyance that some newspapers had indicated that he would "give a full description" at the birthday luncheon, indicating that he was prohibited by financial obligations "involving vast sums of money" from releasing this information. The story concludes, "Before and during the luncheon, Dr. Tesla entertained his guests with colourful personal reminiscences and observations including his opinions of dieting and immortality."

PIGEONS

One of Tesla's biographers, John J. O'Neill, who knew Tesla personally, describes in *Prodigal Genius: The Life of Nikola Tesla* (2006) one of the most remarkable aspects of Tesla's final years – his affection for pigeons. Tesla had been feeding the birds for a long time, but in his final years, his dedication to them seems to have intensified considerably. Initially, he writes, he would be seen "dressed in the height of fashion", but as time went on, the otherwise intensely germophobic Tesla paid less attention to his dress.

The biographer regards Tesla as more than an animal lover, and in fact describes Tesla's relationship with one pigeon as "the world's most fantastic, yet tender and pathetic love affair". To buttress his account, he reports that a science writer on the *New York Times* was also present as Tesla shared his story:

> *I have been feeding pigeons, thousands of them, for years; thousands of them, for who can tell – but there was one pigeon, a beautiful bird, pure white with gray tips on its wings; that one was different. It was female. I would know that pigeon anywhere. No matter where I was, that pigeon would find me; when I wanted her I had only to wish and call her and she would come flying to me. She understood me and I understood her. I loved that pigeon. Yes, I love that pigeon, as a man loves a woman, and she loved me. When she was ill, I knew, and understood; she came to my room and I stayed beside her for days. I nursed her back to health. That pigeon was the joy of my life. If she needed me, nothing else mattered. As long as I had her, there was purpose in my life. Then one night as I was lying in my bed in the dark, solving problems, as usual, she flew in through the open window and stood on my desk. I knew she wanted me; she wanted to tell me something important, so I got up and went to her. As I looked at her I knew she wanted to tell me – she was dying. And then, as I got her message, there came a light from her eyes – powerful beams of light. Yes, it was a real light, a powerful, dazzling, blinding light, a light more intense than I had ever produced by the most powerful lamps in my laboratory. When that pigeon died, something went out of my life. Up to that time I knew with a certainty that I would complete my work, no matter how ambitious my program, but when that something went out of my life I knew my life's work was finished.*

OPPOSITE. A white dove preserved in memory of Nikola Tesla.

RIGHT. The death mask of Tesla on a pedestal, as displayed at the Henry Ford Museum.

TESLA'S DEATH

Tesla's days were indeed numbered. Soon after he claimed to his messenger (see p.71) to have been conversing with the long-dead Mark Twain just the night before, on 8 January 1943, his lifeless body was found in bed by a maid in his New Yorker hotel room. He was 86 years of age. The cause of death was recorded as "coronary thrombosis", perhaps in part because Tesla had recently complained of chest pains.

His funeral services were held at the Cathedral of St. John and included portions in both Serbian and English. Present were numerous diplomats, scientists and engineers, including radio pioneer Edwin Armstrong, as well as over 2,000 other attendees. Nobel laureates Robert Millikan, Arthur Compton and James Franck paid tribute to Tesla as "one of the world's outstanding intellects, who paved the way for many of the important technological developments in modern times".

On the occasion of Tesla's death, First Lady Eleanor Roosevelt sent a message saying, "The President and I are deeply sorry to hear of the death of Mr. Nikola Tesla. We are grateful for his contribution to science and industry and to this country." Meanwhile, Vice President Henry Wallace wrote:

> *Nikola Tesla, Yugoslav born, so lived his life as to make it an outstanding sample of that power which makes the United States not merely an English-speaking nation but a nation with universal appeal. In Nikola Tesla's death the common man loses one of his best friends.*

Perhaps excerpts from the 8 January 1943 *New York Times* obituary best summarize the muddled and sometimes bizarre features of Tesla's later years.

> *Of vigorous temperament and with emphatic ideas on personal health as well as engineering, he had few visitors, according to hotel management, which reported that his meals, strictly vegetarian style, were specially prepared for him by the chef. 'He made everybody keep at a distance greater than three feet,' a hotel executive recalled.*

> *As Tesla advanced in years, his ideas bordered increasingly on the fantastic. In his seventy-seventh birthday interview he had no specific inventions, but said he expected to live 'beyond 140.' The year before, however, he spoke of two great impending discoveries. 'When they are announced,' he said, 'one will be like the 100,000 trumpets of the Apocalypse. The other will be less sensational, but it too will be important. It will be like the shout with which Joshua's army brought down the walls of Jericho.'*

> *He was greatly handicapped by lack of funds, for he was anything but a practical man as far as business was concerned. It was said that he was frequently victimized, but he did not seem to worry much as long as he had a place to work. Tesla probably could have become a rich man had he chosen to become an employee of a large industrial concern, but he preferred poverty and freedom.*

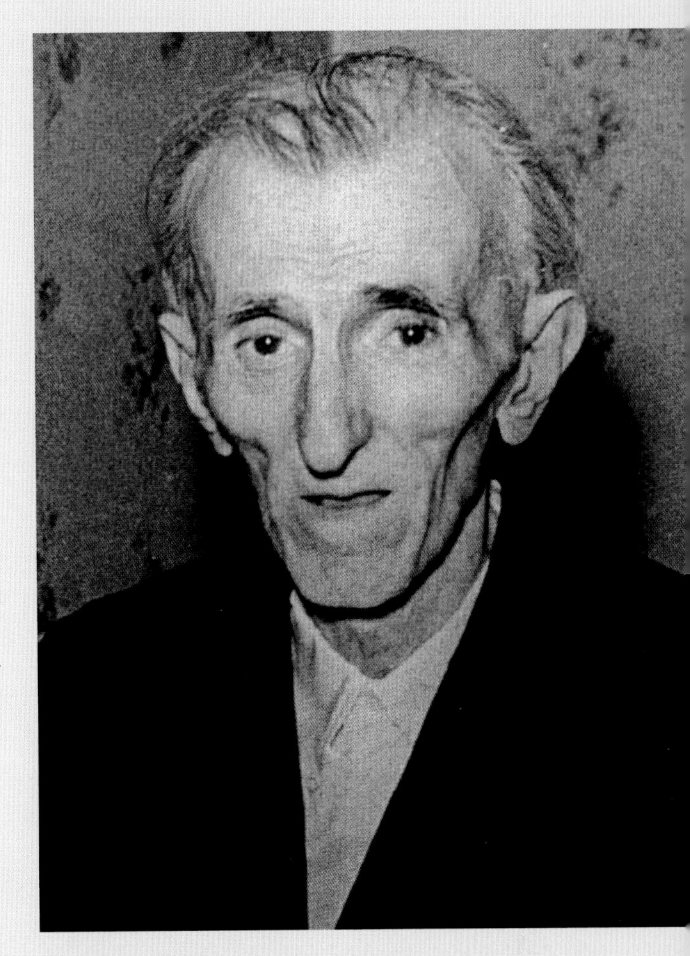

> *Shy of manner and ascetic of tastes, Mr. Tesla preferred his workshop to society. He never married. He ate sparingly and drank neither coffee nor tea because he considered those beverages to be highly injurious. On the other hand, he regarded alcohol in moderation as virtually an elixir of life. It was his habit to stay up until daylight and then sleep only for a few hours before resuming his work.*

ABOVE. The last known photograph of Tesla before his death.

OPPOSITE. First Lady Eleanor Roosevelt visiting US troops. She penned a message praising Tesla on the occasion of his death.

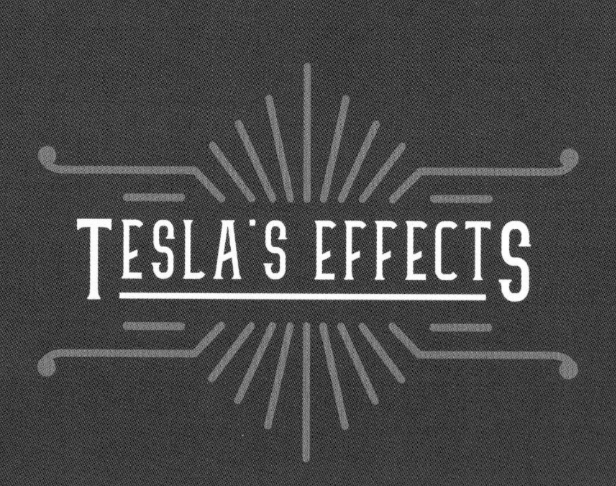

Tesla left no will, and after his death, the Federal Bureau of Investigation took an interest in the disposition of his personal papers and other property. After all, newspaper accounts of Tesla's writings and press conferences had proclaimed the imminent demonstration of a device alternately referred to as a "death ray" or "peace ray", and if such devices or even plans for them had ever been produced, they could have immense implications for national security.

Death-Ray Machine Described

Dr. Tesla Says Two of Four Necessary Pieces of Apparatus Have Been Built.

Amplifying his birthday anniversary announcement of the prospective invention of an electrical death-ray, or force beam, that would make any country impregnable in times of war, Dr. Nikola Tesla says that two of the four pieces of necessary apparatus already have been constructed and tested.

Four machines combine in the production and use of this destructive beam, which, according to Dr. Tesla would wipe out armies, destroy airplanes and level fortresses at a range limited only by the curvature of the earth. These four are:

First, apparatus for producing manifestations of energy in free air instead of in a high vacuum as in the past. This, it is said, has been accomplished.

Second, the development of a mechanism for generating tremendous electrical force. This, too, Dr. Tesla says, has been solved. The power necessary to achieve the predicted results has been estimated at 50,000,000 volts.

Third, a method of intensifying and amplifying the force developed by the second mechanism.

Fourth, a new method for producing a tremendous electrical repelling force. This would be the projector, or gun of the invention.

While the latter two elements in the plan have not yet been constructed, Dr. Tesla speaks of them as practically assured. Owing to the elaborate nature of the machinery involved, he admits it is merely a defense engine, though battleships could be equipped with smaller units and thus armed could sweep the seas.

In addition to the value of this engine for destruction in time of war, Dr. Tesla said it could be utilized in peace for the transmission of power. He had not developed ideas for receiving apparatus capable of transforming the destructive beam into work units, but considered this merely a matter of detail. No suggestion was made of what might happen if an enemy power obtained possession of one of these receiving outfits, and when attacked by the destructive beam simply put it to work in factories manufacturing munitions or uniforms.

Another addition to the anniversary message of the famous inventor was a positive declaration that he expected soon to construct apparatus that would disprove the theories of modern astronomers that the sun gradually was cooling off and eventually the earth would be unable to sustain life, as it would grow too cold.

ABOVE. 1934 report from the *New York Times* describing Tesla's "Death Beam".

OPPOSITE, ABOVE. Three men, including Tesla's relative, Sava Kosanovic (right), in Tesla's New Yorker hotel room after his death.

OPPOSITE, MIDDLE. One of Tesla's many trunks, monogrammed with his initials, part of his estate that was shipped to Belgrade.

OPPOSITE, BELOW. Tesla's estate in the process of being inventoried by staff at the Tesla Museum, Belgrade.

In particular, FBI documents released under the US Government's 1967 Freedom of Information Act reveal that the bureau had its eye on a box that Tesla had placed in the safe at the Governor Clinton Hotel in New York City. Reported to be a "'working model" of an invention, perhaps the rumoured death ray, the report goes on to describe the history of the box's contents:

Placed there by Tesla in 1932 as security for four hundred dollars owed the hotel, this bill is still owed and the hotel appears unwilling to release this property to anyone at least until the debt is paid, but this office will be advised if anyone attempts to pay the bill and obtain the property.

Revealing the bureau's own scepticism that the box could contain any papers or devices that pose a threat to national security, the report goes on to say:

Concerning Tesla, hotel managers report that he was very eccentric if not mentally deranged during the past ten years and it is doubtful if he has created anything of value during that time, although prior to that he probably was a very brilliant inventor. Therefore, any notes of value were probably those made prior to that time.

On the other hand, the bureau was aware of a Yugoslavian ambassador and Tesla relative, Sava

JOHN TRUMP

A professor and electrical engineer at the Massachusetts Institute of Technology named John Trump (uncle of US President Donald Trump) carried out the examination of Tesla's effects, including the box that had been left in the Governor Clinton Hotel safe.

Tesla had warned the hotel staff that no one else should open the box, as its contents were dangerous. It is rumoured that when Trump prepared to open it, others excused themselves from the room. When Trump opened the box, however, he found inside nothing more than an old device for measuring electrical current. Tesla had probably warned against opening the box only to make its contents seem more significant and thereby increase the probability that the hotel would accept it as collateral.

After spending several days reviewing the Tesla effects, Trump concluded that they posed no threat to national security:

> *Tesla's thoughts and efforts during at least the last 15 years were primarily of a speculative, philosophical, and somewhat promotional character often concerned with the production and wireless transmission of power, but did not include new, sound, workable principles or methods for realizing such results. ... It is my considered opinion that there exist among Dr. Tesla's papers and possessions no scientific notes, descriptions of hitherto unrevealed methods or devices, or actual apparatus which constitute a hazard in unfriendly hands. I can therefore see no technical or military reason why further custody of the property should be retained.*

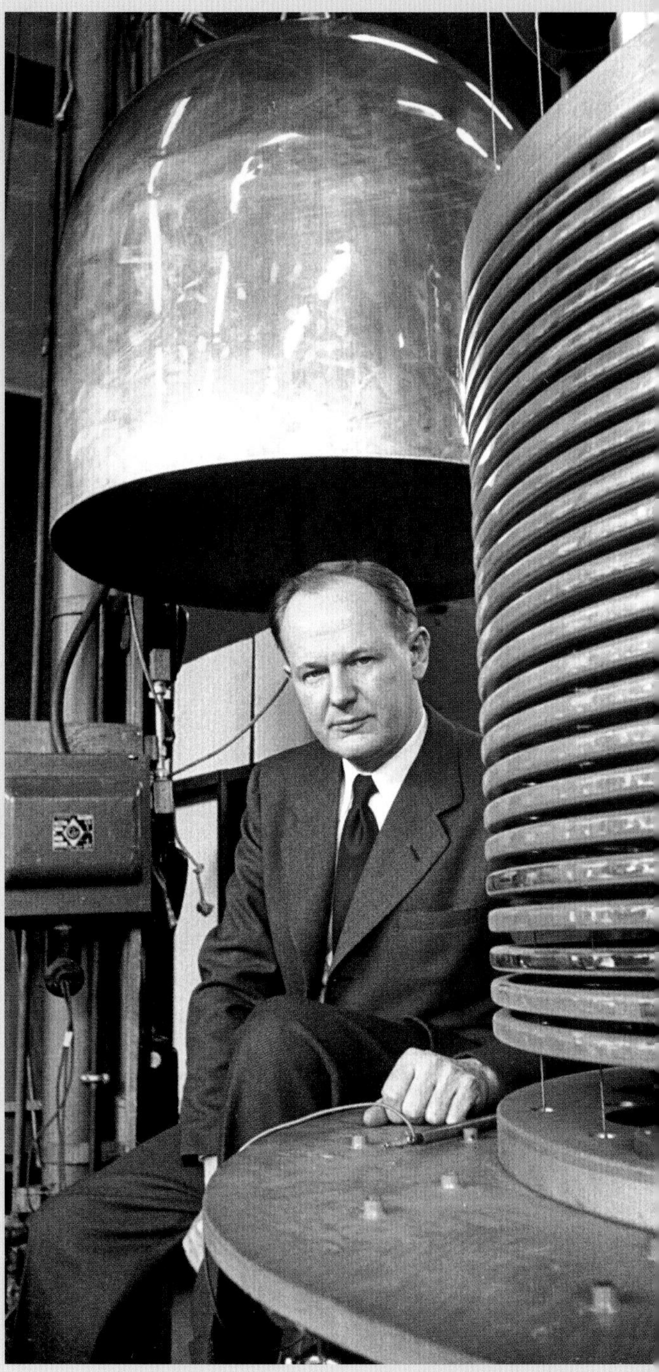

RIGHT. John Trump, MIT physicist who evaluated the significance of Tesla's personal effects after his death.

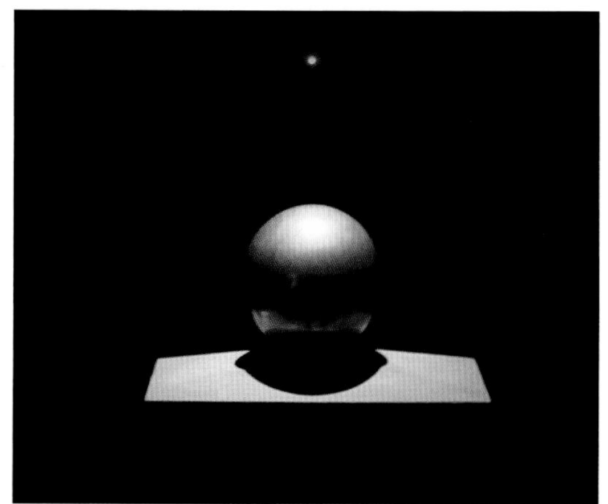

TOP. The Nikola Tesla Museum in Belgrade, Serbia.

ABOVE. Spherical metallic urn containing Tesla's ashes in the Tesla Museum in Belgrade. The sphere was said to be Tesla's favourite shape.

Kosanovic, who was taking steps to take custody of Tesla's papers and effects. Some agency personnel believed "there is a strong likelihood that he will make such information available to the enemy". Records showed that Kosanovic had entered Tesla's hotel room on the day after his death and examined materials in his safe, perhaps even removing some of them.

However, the bureau also reported that, despite the fact that Tesla was a US citizen, the Office of Alien Property's custodian had become involved in the case and had sequestered Tesla's property to permit review by an expert to determine its significance.

Thanks largely to the efforts of Kosanovic, in the 1950s Tesla's effects, totalling 80 trunks, were shipped to Yugoslavia, to be housed in the Tesla Museum in Belgrade. Tesla's ashes were placed in a spherical

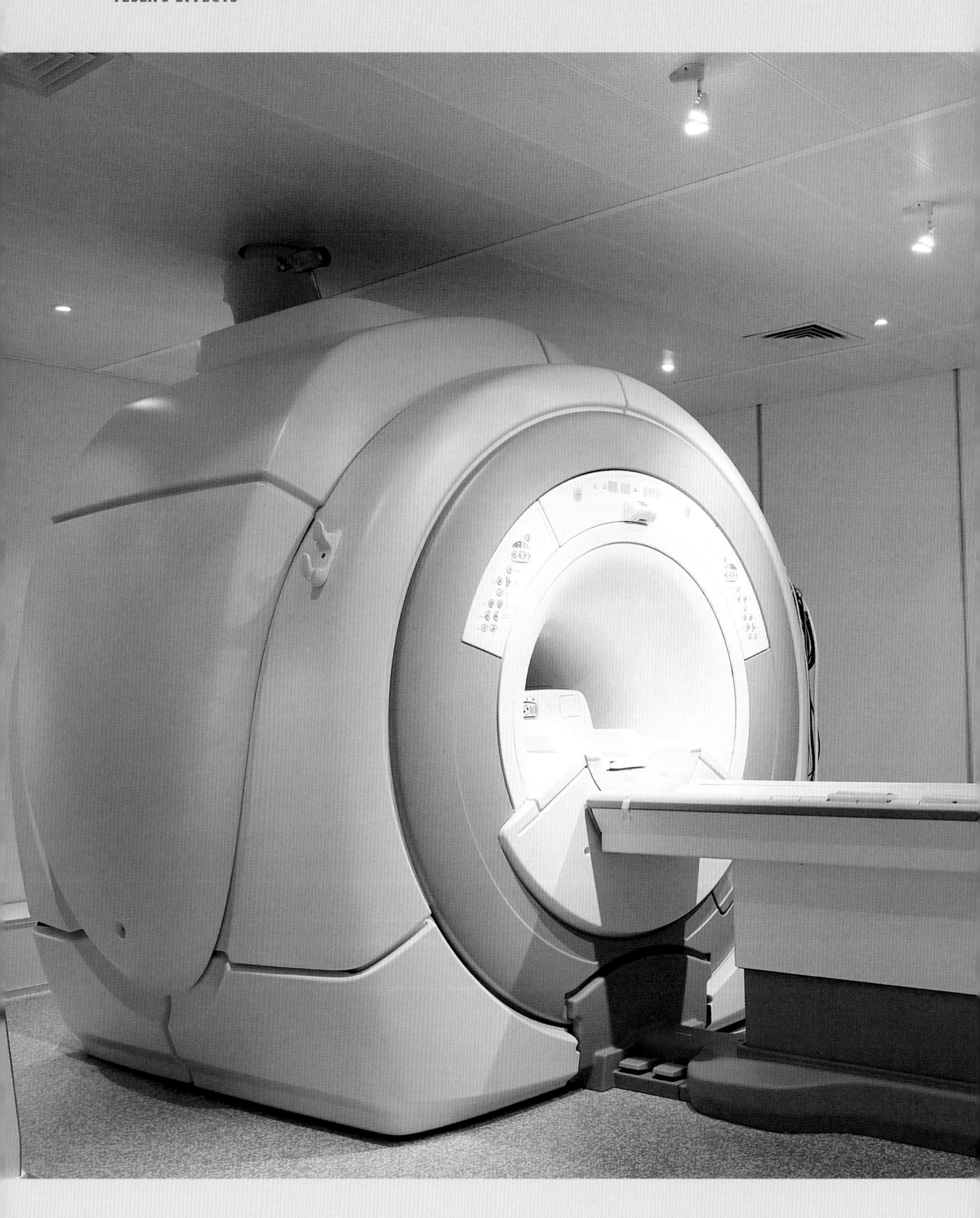

metallic urn. Since then, there have been rumours that various of Tesla's ideas, such as his particle beam weapon, were generating renewed interest among scientists and engineers, but there is no convincing evidence that such devices have been developed or deployed.

In the years since Tesla's death, many rumours have sprung up concerning the nature of his estate. Some claimed that his plans for free worldwide energy were being suppressed by the petroleum industry, and that his particle beam weapon was being kept secret to prevent its development by other nations. Still others claimed that Tesla had left plans or an actual device for communicating with other planets, a "Teslascope", of which they were now in possession.

Others regard Tesla as the greatest prophet of his age, having anticipated a wireless form of telephony and television that people could carry in their pocket – the mobile phone; a wireless technology by which books and even newspapers could be read in the home – the Internet and Wi-Fi; automobiles that could chart their own course through the streets – driverless cars; and aeroplanes capable of taking off and landing both horizontally and vertically – VTOL aircraft.

The scientific community's enduring respect for Tesla is manifest in the 1960 decision of the General Conference on Weights and Measures to name the SI unit of magnetic flux density the Tesla (T). As a radiologist, I regularly make medical diagnoses using magnetic resonance imaging (MRI) scanners whose strength is measured in these units. The most common such MRI strength today is 1.5 T, which is about 30,000 times the strength of the earth's magnetic field at its surface.

LEFT. Photograph of a magnetic resonance imaging (MRI) scanner. Most clinical MRI scanners have a magnetic field strength of 1 to 3 T (Tesla), and the magnet is always "on", even when no patient is being scanned.

NOT AN AXE BUT A HAMMER

In 1962, Abraham Kaplan, then a professor of philosophy at UCLA, formulated what he called the "law of the instrument", which he expressed as "Give a boy a hammer and everything he meets needs to be pounded." Two years later, Kaplan offered a somewhat different formulation: "It comes as no particular surprise to discover that a scientist formulates problems in a way which requires for their solution just those techniques in which he himself is especially skilled."

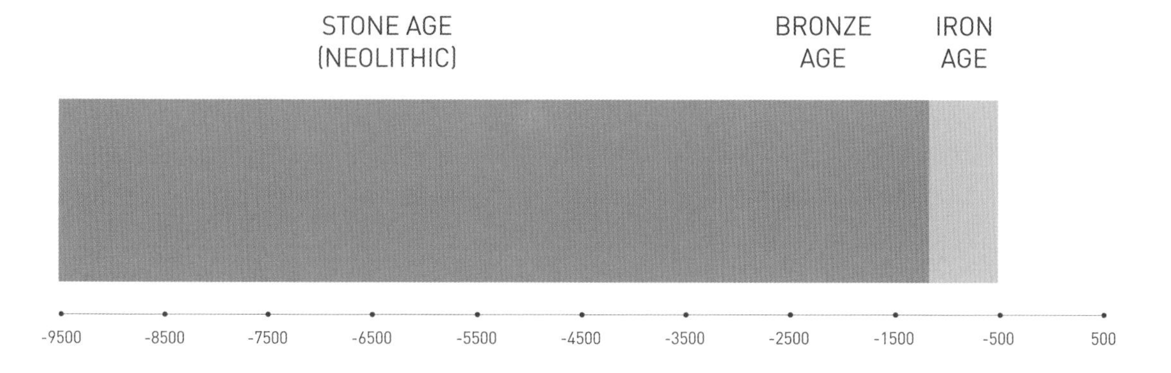

Truer words were never spoken of Nikola Tesla, and they offer deep insight into both his genius and his madness. Let us return to Tesla's 1900 essay, "The Problem of Increasing Human Energy", which offers deep insight into his vision. In it, Tesla conceptualizes human life as a movement, to which "the same general laws of movement which govern throughout the physical universe must be applicable". On this basis, he likens human energy to kinetic energy, according to the formula K.E. = 1/2 m v^2, where K.E. = kinetic energy, m = mass, and v = velocity.

Based on this formula, Tesla deduces that there are only three ways of increasing human energy. The first is to increase m, the mass of mankind. The second is to reduce the forces that tend to reduce human velocity. The third is to increase the forces that tend to propel mankind forward. While the retarding forces can be reduced only to a limited extent (to zero), the impelling force, he argues, can be increased indefinitely.

Tesla says that the mass of mankind can be increased in one of two ways, either by amplifying the forces that tend to increase it or reducing those that diminish it. To achieve this, Tesla advocates careful attention to health, including nutritious food, the promotion of marriage, and good practices in rearing and educating children, with particular attention to the teachings of religion and laws of hygiene, without which the mass will tend to decrease.

Among the programmes that Tesla recommends is the purification of drinking water, which he would accomplish through the electrical production of ozone, the "ideal disinfectant". More broadly, "everyone should recognize his body as a priceless gift, a work of art of indescribable beauty", in relation to which any form of uncleanliness is "not only a self-destructive but highly immoral habit". Above all, Tesla recommends the production of a plentiful supply of healthful food.

To this end, Tesla argues for an end to the "wanton and cruel slaughter of animals", to be replaced by universal vegetarianism. And to increase the yield of the soil, he recommends the production of nitrogen fertilizer though the electrical oxidation of atmospheric nitrogen, a method that can inexpensively increase the fertility of the soil to an almost unlimited degree. Tesla regards the founding of this industry as equal in importance to the dawning of the iron age.

To address the second factor, the reduction in the forces retarding the velocity of mankind, Tesla calls for the reduction of "frictional" forces such as "ignorance, stupidity, and imbecility", as well as the reduction of "negative" forces, foremost among which he ranks "organized warfare". Tesla does not believe that universal peace is yet possible, but he does argue that the number of people engaged in

ABOVE. The first three technological ages of mankind and their approximate time periods. Tesla believed that his innovations would usher in yet another age.

OPPOSITE. Philosopher Abraham Kaplan, who suggested that those in possession of hammers tend to treat all problems as nails.

No. 613.809. Patented Nov. 8, 1898.

N. TESLA.

METHOD OF AND APPARATUS FOR CONTROLLING MECHANISM OF MOVING VESSELS OR VEHICLES.

(No Model.) 5 Sheets—Sheet 5.

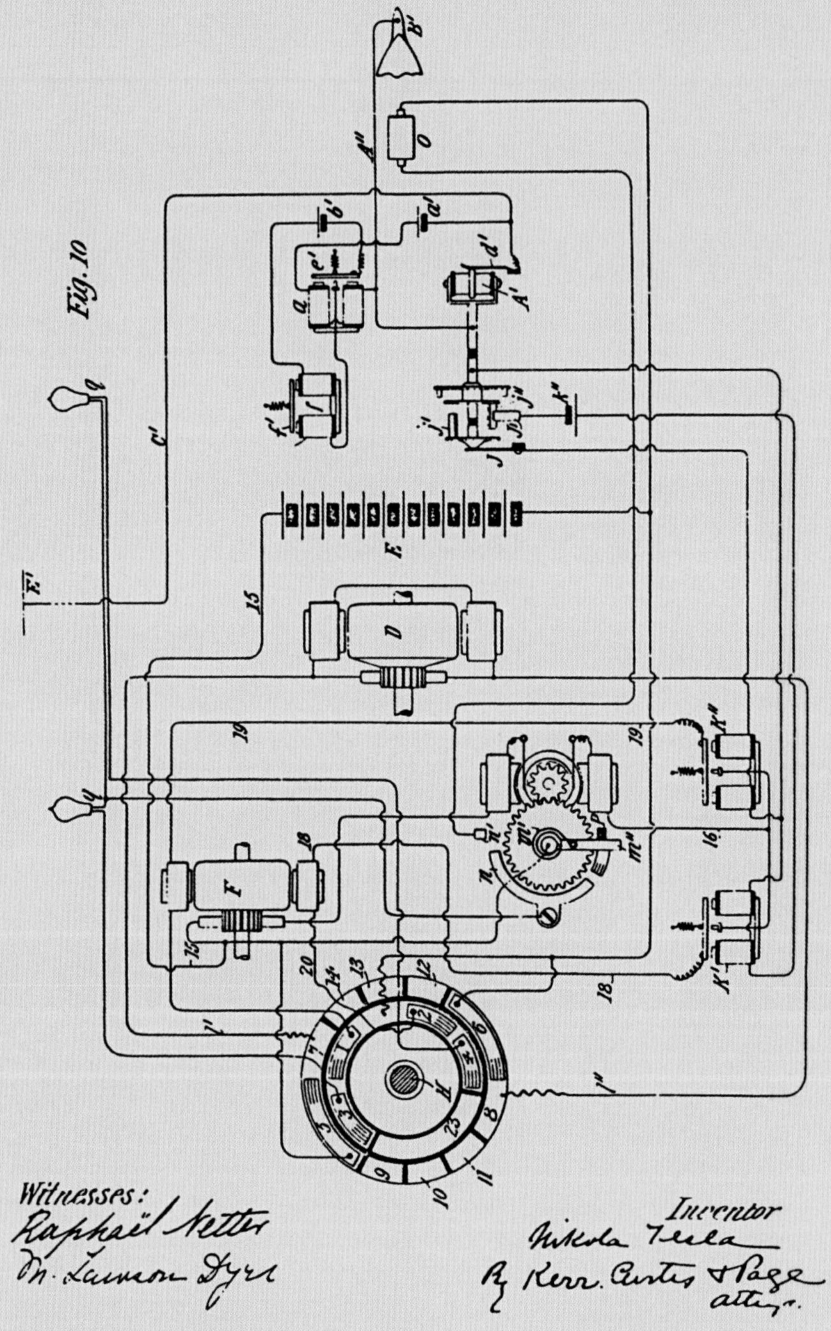

Fig. 10

Witnesses:
Raphaël Netter
M. Lawson Dyer

Inventor
Nikola Tesla
By Kerr, Curtis & Page
Attys.

No. 613,809. Patented Nov. 8, 1898.

N. TESLA.

METHOD OF AND APPARATUS FOR CONTROLLING MECHANISM OF MOVING VESSELS OR VEHICLES.

(No Model.) 5 Sheets—Sheet 1.

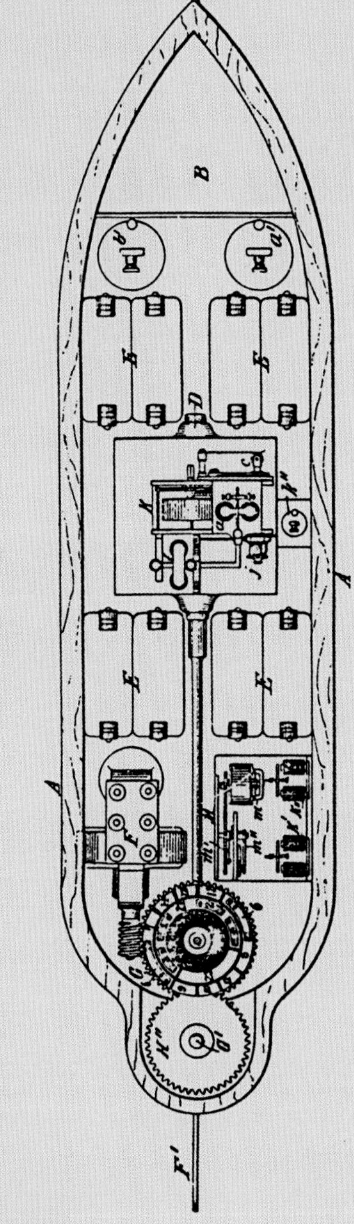

Witnesses:
Raphaël Netter
George Scherff.

Inventor
Nikola Tesla

OPPOSITE AND LEFT.
Diagrams from Tesla's 1898 patent for his "teleautomaton"

bloodshed can be reduced through the application of "teleautomatics", with machine fighting machine.

Tesla offers his radio-controlled boat as the "first practical teleautomaton", envisioning that soon such automata will have their own "minds" and function independently of any operator. In short, warfare would be conducted by fighting machines "without men as a means of attack and defense". Eventually, he believes, such technology will produce a state of "permanent peaceful relations between nations", thereby eliminating the most important force retarding human progress.

As to the problem of increasing the forces propelling mankind forwards, Tesla argues for the harnessing of the Sun's energy. This he would accomplish in three ways: first, by liberating the solar energy stored in coal to produce electricity; second, by using electricity to produce iron more efficiently; and third, by wirelessly transmitting electrical power. Ultimately, however, Tesla envisions that iron will give way to aluminium, saying, "There can be no doubt that the future belongs to aluminum."

Tesla also calls for the development of other sources of power aside from the burning of fuel – namely, what we would call today wind power, solar power and geothermal power. But, he says, none of these would equal in importance the wireless transmission of power, which would do the most to "unite the various elements of humanity", "add to and economize human energy", and "be the best means of increasing the force accelerating the human mass".

Simply put, Tesla's hammer is technology, and his nails – the most pressing problems and greatest opportunities looming before humanity – appear to him technological, as well. Such sentiments are no less common in our own age, when we often tout new technologies – the personal computer, the internet, the cell phone – as the heralds of a new humanity. Yet for worse or better, the transformation of human character cannot be accomplished by the mere expansion of consumer choice.

Of course, in Tesla's mind, the most important tool in humanity's belt is not technology in a generic sense but the technologist par excellence, Nikola Tesla himself. To Tesla, whatever problem he is focused on represents humanity's most significant project. Ultimately, Tesla regards other human beings – including many of the greatest geniuses of his day – as spectators to his own genius, and underlying his grand visions is usually a simple showman's plea: "Look at me!"

ABOVE. Tesla's inventions continued to inspire aspiring scientists throughout the twentieth century, and today.

OPPOSITE. Tesla's achievements were recognized in this letter from the Radio Corporation of America, and in his inclusion in the Smithsonian's first display dedicated to radio.

RADIO CORPORATION
OF AMERICA

RCA RADIO MUSEUM BOARD
G. H. CLARK, SECRETARY

MUSEUM HEADQUARTERS, 75 FRONT STREET, BROOKLYN, N. Y.

TELEPHONES:
FOR OUTSIDE CALLS
FOR INTERDEPARTMENT CALLS

June 30, 1931.

Nikola Tesla, pioneer in electrical arts, and inventor of many of the familiar devices of today, early attempted to solve the transmission of electric power by high-frequency currents - or, as we would say today, by radio.

In 1899, a transmitting station was erected in Colorado Springs, Colorado, to determine experimentally the laws underlying this new inroad into the secrets of electricity. Here Mr. Tesla discovered the presence of stationary electric waves in the earth, and followed this by many other discoveries which led him to believe that "ground waves" were the factors affecting his transmission of high-frequency currents.

Later, a similar tower was located in Waredenclyff, Long Island, and its mushroom-like cap was a familiar, if mysterious, sight to many. Here many further researches were carried on, until, due to a combination of circumstances, experiments ceased.

To dwell upon the many achievements of Nikola Tesla would require volumes. Suffice it here to mention his high-frequency coil, the well-termed "Tesla Coil", which is used today in one form or another at all radio stations. This is still the world's choice when demonstrations of the effect of high voltages are desired.

Under the joint leadership of the Smithsonian Institution and the Radio Corporation of America, plans are being arranged for the housing of a complete engineering and historical record of the birth and growth of radio, in several large centres of the United States, the most complete display to be that of the Smithsonian Institution itself, in Washington, D. C. In this series, the pioneer work of Nikola Tesla will have a prominent position.

G. H. Clark
Curator, RCA Museum Warehouse

PLEASE ADDRESS REPLY TO THIS DEPARTMENT FOR THE ATTENTION OF THE WRITER

THE FLAWED TITAN

A review of Tesla's life and work gives rise to several cautionary insights. One concerns the importance of building and sustaining teams. In contrast to contemporaries such as Edison, Westinghouse and Marconi, Tesla did not manage to build corporations that could carry out his dreams. His method of operation was to patent and then license or sell his ideas, relying on other businesses to manufacture and market the products.

Tesla was, in a word, a loner. To be sure, he had friends, and he was able to sustain a few personal relationships for decades, such as with US diplomat and writer Robert Johnson and his wife Katharine, but by and large his assistants did not remain in his circle longer than a few years. As he grew older and more resentful over his disappointments, his ability to form new relationships faded apace. In the last years of his life, this tendency became still more pronounced, and he died largely a recluse.

This isolation proved problematic, because it deprived Tesla of the opportunity to offset some of his own shortcomings. For example, Tesla tended to believe the brilliance and force of his ideas should be sufficient to carry the day, and he was not particularly interested in putting them to the test. With time, his dreams became further and further detached from the world of practical affairs, to the point that many scientists and engineers no longer took him seriously.

Tesla might have played a much greater role in technology and commerce after the late 1890s had

he worked with a partner or been part of a team that included individuals with a more pragmatic bent, who could have exerted more practical discipline over his highly theoretical tendencies. It is no accident, however, that Tesla attracted no such enduring collaborators, in part because he lived to such a great degree in his own head and could brook little dissent from what he found there.

Tesla recognized the remarkable degree to which his life was in his mind, and the doubts to which it might give rise in others regarding his capacity for useful work. He wrote in 1919, in *My Inventions*:

> *I am credited as being one of the hardest workers, and perhaps I am, if thought is the equivalent of labor, for I have devoted to it almost all of my waking hours. But if work is interpreted to be a definite performance in a specified time according to a rigid rule, then I may be the worst of idlers. Every effort under compulsion demands a sacrifice of life-energy. I never paid such a price. On the contrary, I have thrived on my thoughts.*

Tesla was obsessed with patterns, residing only in hotel rooms with numbers divisible by 3 and insisting on 18 napkins at his table for cleaning his utensils. His disdain for money, as evidenced by his one-sided contract with Morgan, grew directly from his view that only his ideas really mattered. He expected the world around him, and especially the human beings who inhabited it, to conform to his mental models, and when the two did not match, he tended to blame the people.

ABOVE. Katherine Johnson, wife of publisher Robert Johnson and loyal friend of Tesla.

LEFT. Banknote bearing Tesla's image, for 5000 Yugoslav Dinar.

OPPOSITE. Robert Underwood Johnson (1853–1937), American writer and editor of *The Century Magazine*, who, with his wife Katharine, became one of Tesla's most enduring friends.

Tesla seems to have been almost devoid of humility. We might explain it as an unavoidable side effect of genius, but for the fact that other geniuses have proved far more circumspect. He believed, for example, that while his rivals may have secured the present, "the future, for which I really worked, is mine". He anticipates not merely that his discoveries will be vindicated, but that the future of mankind will proceed on a path that only he, Nikola Tesla, has been able to envision.

Increasingly, the later Tesla betrays what can only be called delusions of grandeur, at times not only awkwardly straddling the fine line between genius and madness but dashing straight across it. For example, Tesla attributes the failures of other scientists, engineers and the general public to herald his genius to the fact that he is ahead of his time. If Tesla and the world are out of sync, then humanity had better set about catching up.

On some occasions, Tesla's vision took on a character that, from the present vantage point, can only be described as morally repugnant. For example, in his 1935 essay, "A Machine to End War", Tesla envisions the year 2100, a time by which eugenics will be "universally established":

> *In past ages, the law governing the survival of the fittest roughly weeded out the less desirable strains. Then man's new sense of pity began to interfere with the ruthless workings of nature. As a result, we continue to keep alive and to breed the unfit. The only method compatible with our notions of civilization and the race is to prevent the breeding of the unfit by sterilization and the deliberate guidance of the mating instinct. Several European countries and a number of states of the American Union sterilize the criminal and the insane. This is not sufficient. The trend of opinion among eugenicists is that we must make marriage more difficult. Certainly no one who is not a desirable parent should be permitted to produce progeny. A century from now it will no more occur to a normal person to mate with a person eugenically unfit than to marry a habitual criminal.*

While many sorts of objections might be raised to such a vision, the most salient is Tesla's assumption that it will be possible to distinguish clearly between those who deserve to reproduce and those who do not. On what basis is such a line to be drawn, and who would assume the responsibility for distinguishing the eugenically unfit from the normal? It is ironic that by this point in his life even some of Tesla's most ardent defenders might have questioned his mental fitness.

When it comes to the Cartesian "ghost in the machine", Tesla is all machine and no ghost. As early as 1900's "The Problem of Increasing Human Energy", he describes himself as "an automaton endowed with power of movement, which merely responds to external stimuli beating upon my sense organs and thinks and acts accordingly". So much the engineer, so at home with machines, Tesla found other aspects of reality – including human relationships – beyond his abilities.

ABOVE. A wall plaque in Zagreb, Croatia, commemorating a speech Tesla made in 1892.

OPPOSITE. Tesla in his New York laboratory demonstrating his resonant transmitter.

INDEX

Picture Credits

The publishers would like to thank the following sources for their kind permission to reproduce their photographs in the book:

AKG-Images: Collection Joinville 153b; /Science Source 21, 55b 103

Alamy: Archive PL 133; /Art Collection 19; /Everett Collection Historical 76; /The History Collection 16l; /Niday Picture Library 42; /Pictorial Press 113; /PjrTravel 154; /RTRO 94, 105, 48l; /Science History Images 59, 102; / World History Archive 52

C. D. Arnold: 55t

ESA: 96

Saffron Blaze: 61

Getty Images: BEA: 100; /Bettmann 104; /Alfred Eisenstaedt/The LIFE Picture Collection 142; /Chicago History Museum 56-57; /Denver Post 146; /Encyclopaedia Britannica/UIG 14; /Fotosearch 72, 85t; /PG/Archive Photos 109; /Historic Collection 152; /Hulton Archive 23; /Hulton-Deutsch Collection/Corbis 124; /Philipp Kester/Ullstein Bild 80; /Charles Phelps Cushing/ClassicStock 35; /New York Public Library 28, 29

From the Collections of The Henry For. In Memory of Hugo Gernsback. The Henry Ford: 137

IEEE Awards: 127l

Library of Congress, Washington: 16r, 25, 26, 30, 31, 34, 42, 77, 81, 83, 101t, 101b, 117, 119b,

The Literary Digest: 107t, 107b

National Archives and Records Administration, Washington: 138

NMM: 111

REX/Shutterstock: Associated Newspapers 122; /Marko Drobnjakovic/ AP 143b; /Granger 5, 73, 88b, 99, 112, 114, 115, 129, 135, 150; /Universal History Archive 64

Photo Courtesy USAF: 88t

Public Domain: 9t, 9b, 12, 13b, 24, 27b, 30, 31, 54, 66, 71, 78, 79, 84, 116, 125t, 125b, 130, 139, 148

Science Photo Library: 37; /Ramon Andrade 3Dciencia 128; /Humanities & Social Sciences Library/New York Public Library 85b; /Living Art Enterprises 67; /Peter Menzel 47; /Nikola Tesla Museum 15, 17, 18t, 18b, 18c, 20, 22, 27t, 33, 38, 46, 58, 62, 68-69, 75l, 75r, 86, 93, 97, 106t, 106b, 108, 110, 119l, 119r, 134t, 136, 153t, 155; /Royal Institution of Great Britain 127; /Science Source 39t, 39b, 65b; /Shelia Terry 10, 37t; /SSPL 60, 123; / Universal History Archive 13tl, 44-45

Shutterstock: Yasemin Olgunoz Berber 143t; /Chris Dorney 83b; / Georgios Kollidas 13tr; /Luisrsphoto 50-51; /Nimon 144-145; /Pitk 63

Tesla Collection: 121, 126

Tesla Universe Photos: 32, 40, 41, 53, 68, 90, 91, 95, 120, 131, 134b, 140, 141t, 141c, 141b, 149, 151

Topfoto.co.uk: Fortean 48r, 49

Every effort has been made to acknowledge correctly and contact the source and/or copyright holder of each picture, and Carlton Books apologizes for any unintentional errors or omissions, which will be corrected in future editions of this book.

Tesla, Nikola. *My Inventions: The Autobiography of Nikola Tesla*. New York: Cosimo, 2007.
Martin, Thomas Commerford. *The Inventions, Researches, and Writings of Nikola Tesla*. New York: The Electrical Engineer, 1894..
Carlson, W. Bernard. *Tesla: Inventor of the Electrical Age*. Princeton: Princeton University Press, 2013.
Seifer, Marc. *Wizard: The Life and Times of Nikola Tesla*. New York: Citadel Books, 1998.
Cheney, Margaret. *Tesla: Man Out of Time*. New York: Touchstone, 2011.
Gunderman, Richard; Alavanja Aleks. "Nikola Tesla: An Extraordinary Life". *Radiology* 2015;275:5-8.
Gunderman, Richard. "Nikola Tesla: The Extraordinary Life of a Modern Prometheus". *The Conversation*. January 3, 2018. Available at: https://theconversation.com/nikola-tesla-the-extraordinary-life-of-a-modern-prometheus-89479.
Glenn, Jim. *The Complete Patents of Nikola Tesla*. New York: Barnes and Noble, 1994.
Pavićević, Aleksandra. "From lighting to dust death, funeral and post mortem destiny of Nikola Tesla". Glasnik Etnografskog instituta SANU. 2014;62: 125-139.
Roguin, Ariel. "Historical Note: Nikola Tesla: The man behind the magnetic field unit". J. Magn. Reson. *Imaging* 2004;19:369-374.
Jonnes, Jill. *Empires of Light: Edison, Tesla, Westinghouse, and the Race to Electrify the World*. New York: Random House, 2004.
O'Neill, John. *Prodigal Genius: The Life of Nikola Tesla*. New York: Cosimo, 2006.

Websites
Tesla Memorial Society of New York. www.teslasociety.com.
Nikola Tesla Universe. www.teslauniverse.com.
Nikola Tesla Museum. http://www.yurope.com/org/tesla/arhive.htm.